The Chemistry of Fungi

# The Chemistry of Fungi

**James R. Hanson**
*Department of Chemistry, University of Sussex, Brighton, UK*

RSCPublishing

ISBN: 978-0-85404-136-7

A catalogue record for this book is available from the British Library

Published by The Royal Society of Chemistry,
Thomas Graham House, Science Park, Milton Road,
Cambridge CB4 0WF, UK

Registered Charity Number 207890

For further information see our web site at www.rsc.org

# *Preface*

The diverse structures, biosyntheses and biological activities of fungal metabolites have attracted chemists for many years. This book is an introduction to
the chemistry of fungal metabolites. The aim is to illustrate, within the context
of fungal metabolites, the historical progression from chemical to spectroscopic
methods of structure elucidation, the development in biosynthetic studies from
establishing sequences and mechanisms to chemical enzymology and genetics
and the increasing understanding of the biological roles of natural products.

Fungi occupy an important place in the natural world. As non-photosynthetic
organisms they obtain their nutrients from the degradation of organic material.
They use many of their secondary metabolites to secure a place in a competitive
natural environment and to protect themselves from predation.

Fungi are ubiquitous and their activities affect many aspects of our daily
lives, whether it be as sources of pharmaceuticals and food or as spoilage organisms and the causes of diseases in plants and man. The chemistry of the
fungi involved in these activities has been the subject of considerable study,
particularly over the last 50 years. Although their ramifications can be large, as
in the spread of plant diseases, the quantities of metabolites that could be
isolated precluded much chemical work until the advent of spectroscopic
methods. Whereas many natural products derived from plants were isolated
before the 1960s on a scale that permitted extensive chemical degradation this
was rarely the case for fungal metabolites. However, whenever it was possible,
interesting chemistry was discovered.

This book begins with an historical introduction followed by a description of
the general chemical features that contribute to the growth of fungi. There are
many thousands of fungal metabolites whose structures are known and it is not
the purpose of the book to list them all. There are databases that fulfil this role.
The aim is to describe some of the more important metabolites classified ac-
cording to their biosynthetic origin. Biosynthesis provides a unifying feature
underlying the diverse structures of fungal metabolites. Therefore, the next
chapters begin with a general outline of the relevant biosynthetic pathway

The Chemistry of Fungi
By James R. Hanson
© James R. Hanson, 2008

before presenting a detailed description of particular metabolites. Investigations into these biosyntheses have utilized many subtle isotopic labelling experiments. Compounds that are fungal pigments and those that are distinctive metabolites of the more conspicuous Basidiomycetes are treated separately. Many fungal metabolites are involved in the interactions of fungi with plants and others are toxic to man. Some of these are described in the subsequent chapters. Fungi can transform chemicals in ways that can complement conventional reactions. The use of fungi as reagents forms the subject of the final chapter.

This book owes a great deal to Brian Turner's *Fungal Metabolites*, volumes which cover the literature to 1982, although I have attempted to present more of the chemistry and biological activity than was described in those volumes. A great deal has been discovered since they were published. Reviews on various aspects of microbiological chemistry that have appeared in *Natural Product Reports* and elsewhere are cited in the bibliography.

Finally, I wish to thank Dr Brian Cross and Dr John Grove who first introduced me to microbiological chemistry at The Frythe, Professor Tom Simpson FRS who read the manuscript and Dr Merlin Fox of the Royal Society of Chemistry for his help in the production of the book.

James R. Hanson
University of Sussex

# Contents

**Chapter 1 Fungi and the Development of Microbiological Chemistry**

| | | |
|---|---|---|
| 1.1 | Introduction | 1 |
| 1.2 | Structure of Fungi | 2 |
| 1.3 | Classification of Fungi | 4 |
| 1.4 | The Fungal Cell Wall | 5 |
| 1.5 | History of Fungal Metabolites | 6 |
| | 1.5.1 Fungal Metabolites in the Nineteenth Century | 7 |
| | 1.5.2 Fungal Metabolites 1900–1940 | 8 |
| | 1.5.3 Fungi in the Antibiotic Era, 1940–1960 | 10 |
| | 1.5.4 Study of Fungal Plant Diseases 1940–1965 | 12 |
| | 1.5.5 Impact of Spectroscopic Methods on Structure Elucidation | 13 |
| | 1.5.6 Fungal Metabolites 1965–2005 | 13 |
| | 1.5.7 History of Biosynthetic Studies with Fungi | 15 |

**Chapter 2 The Chemistry of Growing Fungi**

| | | |
|---|---|---|
| 2.1 | The Culture Medium | 18 |
| 2.2 | Laboratory Fermentation | 20 |
| 2.3 | Isolation of Fungal Metabolites | 21 |
| 2.4 | The Stages in a Fermentation | 23 |
| 2.5 | Utilization of the Constituents of the Medium | 25 |
| 2.6 | Fungi Growing in the Wild | 28 |
| 2.7 | Biosynthetic Experiments | 29 |

**Chapter 3 Fungal Metabolites Derived from Amino Acids**

| | | |
|---|---|---|
| 3.1 | Introduction | 32 |
| 3.2 | Penicillins | 33 |

The Chemistry of Fungi
By James R. Hanson
© James R. Hanson, 2008

3.3   Cephalosporins                                              36
3.4   Biosynthesis of β-Lactams                                   37
3.5   Metabolites Containing a Diketopiperazine Ring              39
      3.5.1   Mycelianamide                                       40
      3.5.2   Gliotoxin                                           40
3.6   The Cyclopenin-Viridicatin Group
      of Metabolites                                              42
3.7   Tryptophan-derived Metabolites                              42
3.8   Glutamic Acid Derivatives                                   44
3.9   Fungal Peptides                                             45


**Chapter 4   Polyketides from Fungi**

4.1    Introduction                                               47
4.2    Polyketide Biosynthesis                                    48
4.3    Triketides                                                 50
4.4    Tetraketides                                               51
       4.4.1   6-Methylsalicylic Acid                             51
       4.4.2   Patulin and Penicillic Acid                        52
       4.4.3   Gladiolic Acid and its Relatives                   55
       4.4.4   Tetraketide Tropolones                             56
       4.4.5   Mycophenolic Acid                                  57
4.5    Pentaketides                                               58
       4.5.1   Citrinin                                           58
       4.5.2   Terrein                                            60
4.6    Hepta- and Octaketides                                     61
       4.6.1   Griseofulvin                                       61
       4.6.2   Cladosporin (Asperentin)                           64
4.7    Polyketide Lactones                                        65
4.8    Statins                                                    66
4.9    Cytochalasins                                              68
4.10   Fatty Acids from Fungi                                     68
4.11   Polyacetylenes from the Higher Fungi                       70


**Chapter 5   Terpenoid Fungal Metabolites**

5.1    Introduction                                               73
5.2    Biosynthesis of Fungal Terpenoids                          73
5.3    Monoterpenoids                                             76
5.4    Sesquiterpenoids                                           76
       5.4.1   Cyclonerodiol                                      77
       5.4.2   Helicobasidin                                      78
       5.4.3   Trichothecenes                                     78
       5.4.4   PR-Toxin                                           81
       5.4.5   Botryanes                                          81

|  |  |  |
|---|---|---|
| 5.4.6 | Culmorin and Helminthosporal | 84 |
| 5.4.7 | Sesquiterpenoids of the Basidiomycetes | 85 |
| 5.5 | Diterpenoid Fungal Metabolites | 93 |
| 5.5.1 | Virescenosides | 94 |
| 5.5.2 | Rosanes | 94 |
| 5.5.3 | Gibberellins and Kaurenolides | 97 |
| 5.5.4 | Aphidicolin | 101 |
| 5.5.5 | Pleuromutilin | 102 |
| 5.5.6 | Fusicoccins and Cotylenins | 102 |
| 5.6 | Sesterterpenoids | 104 |
| 5.7 | Fungal Triterpenoids and Steroids | 105 |
| 5.7.1 | Ergosterol | 106 |
| 5.7.2 | Fusidane Steroidal Antibiotics | 107 |
| 5.7.3 | Viridin, Wortmannin and their Relatives | 111 |
| 5.7.4 | Triterpenoids of the Basidiomycetes | 113 |
| 5.8 | Meroterpenoids | 116 |

**Chapter 6  Fungal Metabolites Derived from the Citric Acid Cycle**

| 6.1 | Introduction | 120 |
|---|---|---|
| 6.2 | Citric Acid and Related Acids | 120 |
| 6.3 | Fungal Tetronic Acids | 122 |
| 6.4 | Canadensolide and Avenaciolide | 123 |
| 6.5 | Nonadrides | 124 |
| 6.6 | Squalestatins | 126 |

**Chapter 7  Pigments and Odours of Fungi**

| 7.1 | Introduction | 127 |
|---|---|---|
| 7.2 | Polyketide Fungal Pigments | 128 |
| 7.2.1 | Fumigatin | 128 |
| 7.2.2 | Auroglaucin and Flavoglaucin | 129 |
| 7.2.3 | Hydroxyanthraquinone Pigments | 129 |
| 7.2.4 | Xanthone and Naphthopyrone Pigments | 130 |
| 7.2.5 | Extended and Dimeric Quinones | 131 |
| 7.3 | Fungal Pigments Derived from the Shikimate Pathway | 132 |
| 7.3.1 | Terphenyls | 132 |
| 7.3.2 | Pulvinic Acids | 133 |
| 7.4 | Some Pigments Containing Nitrogen | 135 |
| 7.5 | Terpenoid Pigments | 138 |
| 7.5.1 | Fungal Carotenoids | 138 |
| 7.6 | Lichen Substances | 140 |
| 7.7 | Odours of Fungi | 142 |
| 7.7.1 | Organoleptic Components of Mushrooms | 142 |
| 7.7.2 | Volatile Fungal Metabolites Containing Sulfur | 144 |

**Chapter 8    The Chemistry of Some Fungal Diseases of Plants**

| | | |
|---|---|---|
| 8.1 | Introduction | 147 |
| 8.2 | General Chemistry of Plant–Fungal Interactions | 148 |
| 8.3 | Chemistry of some Leaf-spot Diseases | 149 |
| | 8.3.1 *Botrytis cinerea* | 149 |
| | 8.3.2 *Alternaria* Leaf-spot Diseases | 151 |
| | 8.3.3 *Cercospora* Leaf-spot Diseases | 153 |
| | 8.3.4 Diseases Caused by *Colletotrichum* Species | 154 |
| 8.4 | Fungal Diseases of the Gramineae | 155 |
| 8.5 | Root-infecting Fungi | 157 |
| 8.6 | Some Fungal Diseases of Trees | 159 |
| | 8.6.1 Dutch Elm Disease | 159 |
| | 8.6.2 Eutypa Dieback | 160 |
| | 8.6.3 *Armillaria mellea* | 161 |
| | 8.6.4 *Phytophthora cinnamomi* | 162 |
| | 8.6.5 Silver-leaf Disease | 162 |
| | 8.6.6 *Nectria galligena* Canker | 162 |
| | 8.6.7 Canker Diseases of Cypress | 163 |
| 8.7 | *Trichoderma* Species as Anti-fungal Agents | 163 |
| 8.8 | Fungal Diseases of Plants and Global Warming | 164 |

**Chapter 9    Mycotoxins**

| | | |
|---|---|---|
| 9.1 | Introduction | 165 |
| 9.2 | Ergotism | 165 |
| 9.3 | Trichothecenes as Mycotoxins | 166 |
| 9.4 | Other *Fusarium* Toxins | 168 |
| 9.5 | Aflatoxins | 169 |
| 9.6 | Mycotoxins of *Penicillium* Species | 171 |
| 9.7 | Poisonous Mushrooms | 173 |

**Chapter 10  Fungi as Reagents**

| | | |
|---|---|---|
| 10.1 | Introduction | 177 |
| 10.2 | Xenobiotic Transformations | 177 |
| | 10.2.1 Microbial Hydrolysis | 178 |
| | 10.2.2 Microbial Redox Reactions | 179 |
| | 10.2.3 Microbiological Hydroxylation | 180 |
| 10.3 | Biosynthetically-patterned Biotransformations | 183 |

Contents                                                    xi

**Epilogue**                                                188

**Further Reading and Bibliography**                        190

**Glossary**                                                204

**Subject Index**                                           209

CHAPTER 1

# Fungi and the Development of Microbiological Chemistry

## 1.1 Introduction

Fungi are widespread, non-photosynthetic microorganisms that play a vital role in the environment, particularly in the biodegradation of organic material. The study of their metabolites and metabolism has made many contributions to the overall development of chemistry. Although the biosynthetic pathways that fungi utilize to construct their metabolites have general features in common with those found in bacteria, plants and mammals, they differ in detail and the structures of the resultant natural products are often different. This book is restricted to fungal metabolites but the reader should not lose sight of other natural products produced elsewhere in the living world.

Since fungi do not contain chlorophyll and are not photosynthetic organisms, they gain their energy and many of the nutrients to supply their biosynthetic pathways through the degradation of plant and other matter. Their environmental role is that of recycling. Their widespread provenance is often illustrated in one of the first practical exercises of many microbiology courses. A Petri dish containing a nutrient agar is exposed to the atmosphere for a few minutes. It is then incubated to reveal the range of organisms, both bacteria and fungi, whose spores are present in the atmosphere and which fell onto the plate in a relatively short time. It was a chance contaminant of an agar plate that led to the isolation of penicillin and changed the face of medicinal chemistry.

Fungi are eukaryotic organisms with a distinct nucleus, unlike bacteria which are prokaryotes. This also distinguishes them from another wide family of soil microorganisms, the Actinomycetes (*e.g.* Streptomycetes), which are often considered along with the bacteria. Yeasts, however, are regarded as a unicellular form of a fungus. Some fungi grow in a symbiotic relationship with photosynthetic algae or cyanobacteria in the form of lichens.

The Chemistry of Fungi
By James R. Hanson
© James R. Hanson, 2008

Fungi do not grow in isolation. Some attack plants, insects and mammals as pathogens whilst others are saprophytic and grow on dead material. Some live in a positive symbiotic relationship with a host organism. Thus, there are mycorrhizal fungi that are associated with the roots of plants and facilitate the uptake of nutrients by the plant. Others are endophytic organisms that grow within the vascular system of the plant. Throughout the natural world there is a chemical language between the fungus and its host which determines the nature of this relationship. We are beginning to understand the role of fungal and plant metabolites in this ecological communication.

The chemistry of fungi impinges on many aspects of our daily life whether it be in the role of yeasts in the production of bread and wine, the edible mushrooms or the manufacture of antibiotics such as the penicillins. The fungal diseases of crops, ornamental plants and trees and the spoilage of stored foodstuffs are serious economic problems. The control of the phytopathogenic organisms and the detection of their toxic metabolites in the food chain provide further chemical problems.

The microbiological chemist is interested in the structure, chemistry and biological activity of fungal metabolites. The biosynthesis of these metabolites, the sequences, stereochemistry and mechanism of the individual steps, together with the structure and regulation of the enzymes involved, is a major area of enquiry. The ecological chemistry of fungal interactions with plants and insects has provided another area of chemical investigation. An understanding of the chemical basis of fungal bio-control agents may have useful agrochemical applications.

As biodegradative organisms, fungi can carry out microbiological transformations of extraneous chemical substances. They can behave as self-replicating, environmentally friendly, chiral reagents. Their ability to carry out transformations that are chemically difficult, *e.g.* hydroxylations at sites that are remote from other reactive centres, has been exploited commercially. The scope of these biotransformations and the development of predictive models so that the use of an organism can be built into a synthetic strategy is yet another area of investigation. The use of the biodegradative ability of fungi in the bio-remediation of contaminated land is a further application of chemical interest.

There are various estimates of the number of species of fungi. These range from 100 000 to 250 000. What is clear is that only a relatively small number, of the order of a few thousand, have been thoroughly investigated by microbiological chemists. Furthermore, there are often different strains of the same species. Whilst these may be morphologically similar, their metabolites can be quite diverse. Some metabolites may be produced consistently by all the strains of a particular species whilst other metabolites may be variable. The chemistry of an organism can also vary with the conditions under which it is grown. Unsurprisingly, therefore, some species of economic importance, *e.g. Penicillium chrysogenum*, have generated immense chemical interest.

## 1.2   Structure of Fungi

At first sight the structures of fungi appear quite diverse. The fruiting body of the common edible mushroom, *Agaricus bisporus*, is very different from the green *Penicillium* species growing on the surface of some cheese. However, there are some common features. The basic structural units of most fungi are the filaments known as the hyphae. Collectively, hyphae can aggregate to form a felt known as the mycelium. In some of the higher fungi, the hyphae can aggregate to form long strands and even differentiate to create a structure almost like a boot-lace, which is known as a rhizomorph. Another name for the honey-fungus, *Armillaria mellea*, which does considerable damage to trees, is the 'boot-lace fungus', which aptly describes the rhizomorphs by which it spreads underground.

The higher fungi, the mushrooms and toadstools, develop complex and readily observable structures known as fruiting bodies. These sprout from their mycelium, particularly in the autumn, and produce spores. At the other extreme some unicellular micro-fungi, such as the yeasts, produce small globular or ellipsoid cells that are only visible under the microscope.

The hyphae may be long single multi-nucleate aseptate (undivided) cells through which the cellular cytoplasmic fluids may flow. Other hyphae are septate and have distinct divisions. In these much of the chemical activity takes place at the growing tip. The lower micro-fungi only become septate as the culture ages whilst the higher macro-fungi become septate at an early stage and, as rhizomorphs are formed, their function may differentiate.

The form a fungus takes can depend on the culture conditions. Some fungi will have a yeast-like form under one set of conditions and a filamentous form under others. Under inhospitable conditions, often exploited in the storage of cultures, an organism can develop a 'resting' stage. In the wild this can allow spores to over-winter in the soil. In the laboratory, fungal cultures are often stored at low temperatures on agar under oil or in sealed vials on sand.

When a fungus is grown in suspension in a nutrient medium contained within a conical flask, the mycelium will sometimes clump together whilst at other times a well-dispersed mycelial suspension or even a mycelial mat is formed. The aeration and hence the metabolic capabilities of these forms can differ. The aeration can be quite poor within tightly formed clumps and this can affect the metabolism of the fungus. It is often difficult to get higher fungi to produce fruiting bodies in laboratory culture and again this can affect their metabolite production. Some rapidly growing fungi such as *Rhizopus* species produce fine long hyphae that spread rapidly across the agar in a Petri dish. They may produce a covering of aerial mycelium with the appearance of household dust. Indeed, quite a lot of household dust is fungal mycelium.

Fungi usually reproduce by spores although they can also develop vegetatively from mycelial fragments. The spores may be pigmented and some may have a gelatinous polysaccharide coating to facilitate their dissemination by a carrier and their attachment to a host. They are often borne on a specific thallus or germ tube. Hyphae that carry these are known as conidiophores. A culture such as that of *Botrytis cinerea* may appear light grey as the mycelium spreads across a Petri dish and then it develops a ring of green-black sclerotial mycelium bearing spores.

# 1.3   Classification of Fungi

In the general taxonomic classification, fungi are grouped in terms of the following ranks: division, class, order, family, tribe, genus, section, and species. In the binomial description of a fungus, the first name is that of the genus and the second name is the species. The name (not italicized) that follows this may be that of the author who first described the species. There are often varieties and strains of particular species. The accession number in a culture collection can be important in defining the organism used to isolate a particular metabolite. Although some metabolites may be specific to particular strains, others may be more common and are found in a section of a genus. The structure of the reproductive organs and the mechanisms of reproduction form the basis of the classification of fungi. These organisms may be broadly grouped in the following way. There are the Phycomycetes or lower fungi, which have a simple thallus bearing the spores. They possess unicellular aseptate hyphae. In some classifications this class name is treated as a trivial name for the Mastigomycotina and Zygomycotina. Typical examples are the Peronosporales, which include plant pathogens such as *Pythium* and *Phytophthora* species and the Mucorales, which include the common *Mucor, Rhizopus* and *Phycomyces* species. The 'damping-off' fungus *Pythium ultimum*, found growing across over-zealously watered germinating seeds, is an example.

A second group are the higher fungi which have septate hyphae, and these can be divided into the Ascomycetes and the Basidiomycetes. In the Ascomycetes the spores are borne in a sac-like structure known as an ascus. This type of fruiting body or ascocarp is found in *Monascus* species. The fungus *M. ruber*, which produces the red colour on Chinese red rice, is an example of these. The genera *Penicillium* and *Aspergillus* belong to the class of Ascomycetes known as the Plectomycetes. The spores are held in a pear-shaped perithecium in another class known as the Pyrenomycetes. The saprophytic plant parasites of the Hypocreales are also members of this group. Some of the best known of the higher fungi are Basidiomycetes. Here the spores are borne in special distinctive fruiting bodies. The edible part of the common mushroom, *Agaricus bisporus*, is a typical example.

The final large group are the Fungi Imperfecti or Deuteromycetes. In these organisms, the perfect stage of reproduction is rare or unknown and for the most part they are cultured vegetatively. The *Fusaria* are the best known of these. This classification can be confusing because some fungi originally classified within the Fungi Imperfecti do have both an asexual imperfect stage and a perfect stage. Thus the fungus that produces the gibberellin plant hormones, *Gibberella fujikuroi*, is the perfect stage of *Fusarium monoliforme*.

The naming of fungi has undergone many changes over the years and this can be a source of confusion. For example, *Ophiobolus graminis* was the name given to a serious pathogen of wheat. This name was incorporated into that given to a family of terpenoid metabolites, the ophiobolanes, which were isolated from the fungus. However, the fungus is now known as *Gaeumannomyces graminis*. Ophiobolanes are also produced by a rice pathogen that was at one time known as *Helminthosporium oryzae* or *Drechslera oryzae* and is now described as *Bipolaris oryzae*. Many of the *Polyporus* species, which gave their names to the triterpenoid polyporenic acids, have also been renamed as *Pipto-pterus*. When attempting to re-isolate a fungal metabolite, particularly from a culture that has been deposited in one of the culture collections, it is helpful to trace the provenance and naming of a particular isolate. When describing the isolation of a fungal metabolite, it is important to record the accession number of the culture in one of the major culture collections. If the strain of the organism is a new isolate it should be deposited in an accessible culture collection. Much valuable time has been wasted in unsuccessful attempts to re-isolate a fungal metabolite when the original culture has been lost.

## 1.4 The Fungal Cell Wall

The chemistry of the fungal cell wall contains some useful taxonomic markers. The cell wall is also a very important target for anti-fungal agents. The fungal cell wall differs in its structural components both from the bacterial cell wall and mammalian cell membranes. The fungal cell wall is a complex of chitin [a polymer of *N*-acetylglucosamine (**1.1**)], various mannoproteins together with α- and β-linked 1,3-D-glucans. Electron microscopy of the cell walls of the yeast *Candida albicans* shows that they are in layers attached to a plasma membrane. The major sterol in these is ergosterol (**1.2**) rather than cholesterol which is found in mammalian systems. Inhibitors of the biosynthesis of these components can, therefore, be selectively fungicidal.

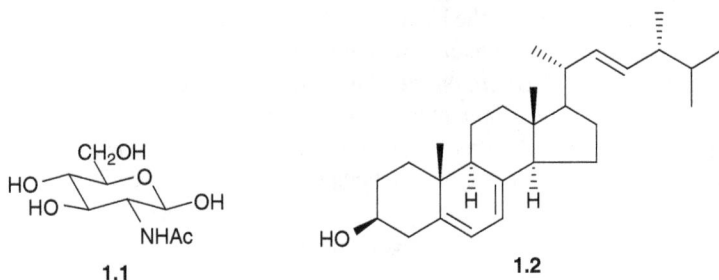

**1.1**     **1.2**

The development of novel anti-fungal agents is a continuing area of research. Furthermore, opportunistic fungal infections, particularly caused by *Candida* and *Aspergillus* species, are emerging as a source of morbidity and mortality amongst immunocompromised patients. Polyene antibiotics such as nystatin and amphotericin B bind to ergosterol much more avidly than to cholesterol and hence disrupt the fungal cell membrane. Ergosterol biosynthesis inhibitors such as the azole fungicides target a key stage in the biosynthesis of ergosterol, the C-14α demethylation of lanosterol. Melanins are dark brown or black pigments that are present in fungal cell walls and are formed by the oxidation of phenolic precursors such as tyrosine (**1.3**), 3,4-dihydroxyphenylalanine (**1.4**) and 1,8-dihydroxynaphthalene (**1.5**). Some anti-fungal agents such as tricyclazole produce a weakening of cell walls by inhibiting melanin biosynthesis. More recently, several compounds that target the β-(1,3)-D-glucan and chitin synthases have been developed.

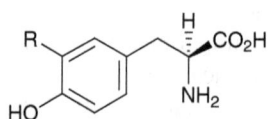

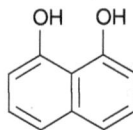

**1.3** R = H          **1.5**
**1.4** R = OH

Part of the antagonistic interaction between fungi, such as that between *Trichoderma* and other organisms, includes the production of a chitinase. This allows the *Trichoderma* to attack the cell wall of its target organism. The hyphae of the *Trichoderma* can then penetrate the target fungus and sequester its nutrients.

## 1.5  History of Fungal Metabolites

The chemical activities of fungi have a long history. Many fungi, because of the competitive environment in which they live, produce antibiotics of varying efficiency. The Greek physician, Dioscorides described the use of an infusion that he called Agaricium, which was obtained from the larch polypore, *Fomitopsis* (*Polyporus*) *officinalis*, and was used for the treatment of consumption (tuberculosis). This biological activity has been attributed to the presence of agaricic or laricic acid [α-cetylcitric acid (**1.6**)]. The 'ice-man', whose 5300 year old body was discovered some years ago in the ice in the Alps between Italy and Austria, had the birch polypore, *Piptoporus betulinus*, with him. This fungus is active against wound bacteria such as *Staphylococcus aureus*. There are records of the use of other fungi, particularly *Ganoderma lucidum*, in ancient Chinese medicine. The identification of moulds growing on cloth by their pigmentation and their treatment is described in the Old Testament of the Bible in Leviticus Chapter 13, Verse 47.

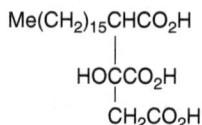

**1.6**

The hallucinogenic properties of fungi such as *Amanita muscari* were known to several peoples. It is possibly the Soma which was used in parts of Asia and Scandinavia. There are records from travellers in the 18th century of its use.

The toxicity of ergot was apparently known to the Syrians in 600 BC. The metabolites of the ergot fungus, *Claviceps purpurea* which grows on rye, contaminated rye bread and brought about the disease known in the Middle Ages as St Anthony's Fire. Ergotism involved damage to the nervous system and vascular constriction, leading to death of the affected parts of the body. Subsequently, medicinal uses of ergot were developed. In the early nineteenth century ergot was used to induce childbirth and to prevent post-natal haemorrhage.

## 1.5.1   Fungal Metabolites in the Nineteenth Century

The fungus *Penicillium glaucum* was used by Pasteur in 1860 to degrade one enantiomer of tartaric acid and allow the resolution of a racemate in experiments that laid the foundations for the study of chirality. Pasteur was also one of the first to recognize the antagonism between microorganisms, which led in 1889 to the use of the term 'antibiote' by a French biologist Vuillemin to describe the substances involved. The term antibiotic was redefined much later by Waksman in 1941 to describe a natural product formed by a microorganism that inhibited the growth or killed another microorganism.

Fungi were recognized as the cause of several serious plant diseases in the middle of the 19th century. The need to control fungal diseases of plants, such as *Plasmopara viticola* infections of vines, led to the development by Millardet in 1883 of Bordeaux mixture (copper sulfate and lime).

The scale on which many natural product degradations were carried out in the late 19th century and the early 20th century precluded much structural work on fungal metabolites. With a few exceptions, fungal material was not available on a scale that would permit the isolation of the gram quantities of natural product that were used for structural studies in the days preceding spectroscopic methods.

One of the exceptions was ergot from *Claviceps purpurea*. The therapeutic as well as the toxic properties of ergot provided the stimulus for studies. Although an impure alkaloidal fraction of ergot had been described by Wenzell in 1866, a crystalline alkaloid, ergotinine, was isolated by Tanret in 1875. However, the structures of these alkaloids derived from lysergic acid (**1.7**) were not elucidated until 1935 when the work of Jacobs and Craig in New York and by Smith and Timons at the Wellcome Laboratories came to fruition. A crystalline yellow pigment, sclererythrin, was first isolated from ergot by Dragendorff in 1877. Subsequent studies led to the isolation of ergoflavin by Freeborn in 1912 and to ergochrysin in 1931 by Barger and Bergmann. The structures of these pigments (*e.g.* **1.8**), which differ from those of the alkaloids, were not established until the late 1950s and early 1960s. The ubiquitous fungal sterol ergosterol was first isolated by Tanret in 1879. The structure of ergosterol (**1.2**) was eventually established in 1933 by Chuang through an inter-relationship of the parent hydrocarbon ergostane with the cholic acids and cholesterol.

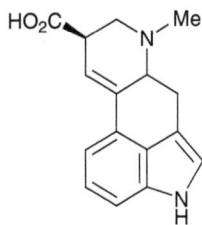

**1.7**

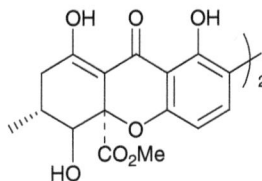

**1.8**

Several simple acids, such as oxalic and citric, were isolated from *Aspergillus niger* in 1891–1893. Commercial methods for the microbiological production of citric acid (**1.9**) were developed in the early 1920s.

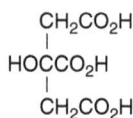

CH$_2$CO$_2$H
|
HOCCO$_2$H
|
CH$_2$CO$_2$H

**1.9**

## 1.5.2   Fungal Metabolites 1900–1940

Mycophenolic acid, which is now used as an immunosuppressant to prevent organ rejection after transplant operations, was first isolated by Gosic from a *Penicillium* species in 1896. It was subsequently isolated from *P. stoloniferum* by Alsberg and Black in 1913 and from *P. brevicompactum* in 1932 by Raistrick. However, its structure (**1.10**) was not finally established until 1952. The much simpler fumaric acid was isolated by Ehrlich in 1911 from *Mucor stolonifer*. The fungus *Aspergillus oryzae* is used in Japan to produce the koji fermentation of rice to make saké. Kojic acid, 5-hydroxy-2-hydroxymethyl-4-pyrone (**1.11**), was first isolated from this organism in 1907 and its structure was established by Yabuta in 1924. The anti-bacterial agent penicillic acid was isolated from *Penicillium puberulum* by Alsberg and Black in 1910. Its structure (**1.12**) was established in 1936 by Raistrick, who had obtained it from *P. cyclopium*.

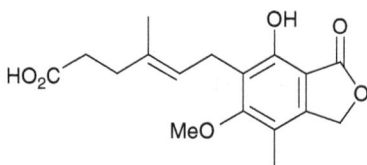
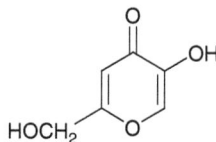
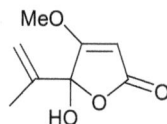

**1.10**                                    **1.11**                                    **1.12**

The First World War saw the development of the bacterial fermentation of *Clostridium acetobutylicum* for the production of acetone and butanol. This provided a stimulus to microbiological chemistry. During the 1920s there were several studies on the effect of the composition of the medium on metabolite production, *e.g.* by *Aspergillus niger*, leading to the isolation of D-gluconic acid lactone (**1.13**) by Malliard in 1923. This compound had previously been obtained by Herrick and May from *Penicillium luteum* in 1912. Commercial fermentation methods using a chiral acetoin condensation mediated by the yeast, *Saccharomyces cerevisiae*, were developed in 1930 for the synthesis of L-(−)-ephedrine from benzaldehyde.

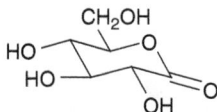

**1.13**

During the 19th and early 20th centuries there had been several reports of poisoning arising from mouldy wall paper. In a study of microbiological methylation, the formation of the poisonous and volatile trimethylarsine through the metabolism of arsenites present as pigments in wall paper by *Scopulariopsis brevicaulis* was established by Challenger in 1933.

Several quinonoid pigments of the higher fungi were identified in the 1920s by Kogl. These included frangulaemodin (**1.14**) from the blood-red agaric *Dermocybe sanguinea* in 1925, polyporic acid (**1.15**) from *Polyporus nidulans* in 1926 and atromentin (**1.16**) from *Paxillus atromentosus* in 1928. The terphenyl thelephoric acid (**1.17**) is a widespread pigment of the Basidiomycetes.

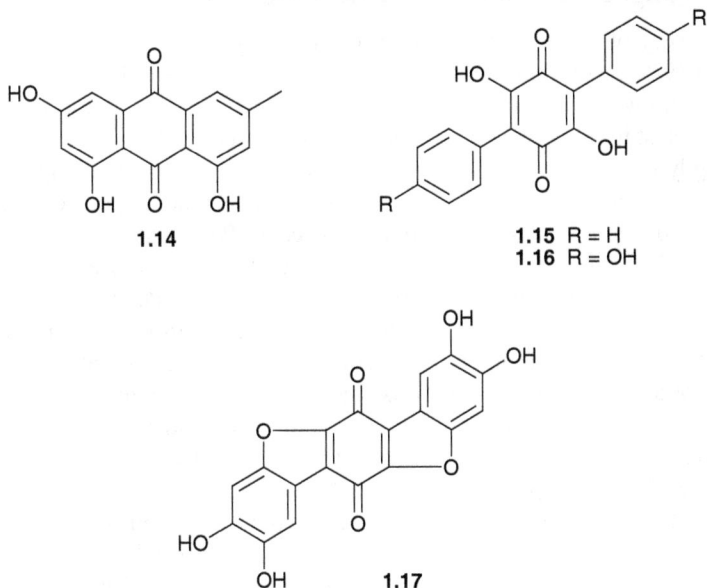

1.14                                  1.15 R = H
                                      1.16 R = OH

1.17

Raistrick studied the consumption of glucose from their growth medium by fungi and concluded that this was a useful guide to the production of fungal metabolites. By monitoring this to establish the period of a fermentation he was able to isolate and identify a series of mainly anthraquinone pigments from various *Penicillium*, *Aspergillus* and *Helminthosporium* species. Much of this work, which started in the Nobel Division of ICI at Ardeer in Ayrshire and was continued at the London School of Hygiene and Tropical Medicine, was published in the 1930s and 1940s.

The constituents of various *Penicillium* species growing as spoilage organisms on maize were investigated at this time in the context of a possible link with pellagra. Although this disease turned out to have a completely different origin, nevertheless the studies led to the identification and elucidation of the structure of several metabolites, including citrinin (**1.18**) from *Penicillium citrinum*, penicillic acid (**1.12**) from *P. cyclopium* in 1936 and a series of tetronic acids such as carlosic acid (**1.19**) from *P. charlesii* in 1934. Another metabolite, puberulic acid, isolated from *P. puberulum* in 1932 and a relative stipitatic acid isolated in 1942 from *P. stipitatum* were assigned theoretically interesting pseudoaromatic tropolone structures by Dewar in 1945.

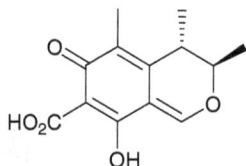

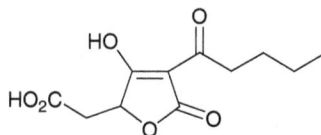

**1.18**                                                    **1.19**

### 1.5.3   Fungi in the Antibiotic Era, 1940–1960

The discovery of penicillin by Fleming in the autumn of 1928, and which he reported in 1929, revolutionized medicinal chemistry after a lengthy gestation period. Fleming who was Professor of Bacteriology at St Mary's Hospital in Paddington, had an interest in the control of the bacterial infection of wounds by various agents, including lysozyme. During August 1928 he left a Petri dish containing a bacterial culture of *Staphylococcus aureus* for washing up. This plate became contaminated by fungal spores of a *Penicillium* species. Whereas *Penicillium* species grow quite well at 18–25 °C, most bacteria grow better at a higher temperature of 30–38 °C. At first, the weather that August was cool, allowing the *Penicillium* to develop. Subsequently, there was a hot period and the bacteria began to develop. However, there were clear zones of inhibition around the fungal contaminant, indicating the presence of an antibiotic. Fleming was able to isolate the fungus and demonstrate that it produced a powerful antibiotic. He originally identified the fungus as *Penicillium rubrum* but this was subsequently corrected by Raistrick and Thom to *P. notatum*. Fleming approached Raistrick for help in isolating the metabolite. However, the instability of penicillin precluded its isolation at that time. The isolation and structural work was taken up by Florey, Chain, Abraham and Heatley in 1938 in Oxford. The outbreak of the Second World War increased the urgency with which the research was undertaken. By 1940 the yield of penicillin had been increased and sufficient material was available for the first human trial to take place in February 1941. Collaboration with Oxford chemists (Robinson, Wilson Baker and Cornforth) and with the Northern Regional Research Laboratories in Peoria in the USA began in 1942/3. The β-lactam structure of penicillin (**1.20**) was eventually established in 1945 as a result of the chemical work and X-ray crystallographic studies by Dorothy Hodgkin. Further details of the work on penicillin are discussed later in the chapter on fungal metabolites derived from amino acids.

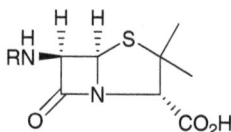

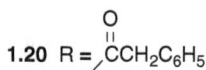

**1.21** R = H

There are, however, several milestones in the development of the penicillins that can be recorded here. The core of the penicillin structure, 6-aminopenicillanic acid (**1.21**), was isolated in 1959, paving the way for the preparation of semi-synthetic penicillins with enhanced biological activity. In 1960 the tripeptide, δ-(L-α-aminoadipoyl)-L-cysteinyl-D-valine (LLD-ACV) was isolated by Arnstein. This peptide has played a central role in biosynthetic studies, particularly by Abraham and Baldwin in Oxford, which have culminated in studies on the enzyme system isopenicillin N synthase and the demonstration of the role of this non-haem iron oxidase in the formation of the penicillin ring system. These studies have continued to the present time. The problem of the resistance of bacteria to the penicillins was already known in the 1940s. The discovery in 1976 of the β-lactamase inhibitor clavulanic acid (**1.22**), which is a metabolite of *Streptomyces clavuligenus*, made a significant contribution to reducing but not eliminating this problem. Methicillin resistant strains of *Staphylococcus aureus* (MRSA) pose a serious medical problem today.

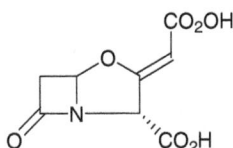

**1.22**

The anti-bacterial activity of the penicillins stimulated a major search for other antibiotics. Many compounds were isolated from amongst the metabolites of *Streptomycetes* obtained from soil. These included therapeutically useful antibiotics such as chloramphenicol, the tetracyclines, erythromycin and streptomycin. Several fungi yielded useful compounds. One of these was *Cephalosporium acremonium*. A biologically-active strain was obtained by Brotzu from a sewage outfall near Cagliari in Sardinia in 1945 and described in 1946. The β-lactam cephalosporin C (**1.23**) was isolated from this organism in 1954 and its structure determined in 1961 by Abraham and Newton.

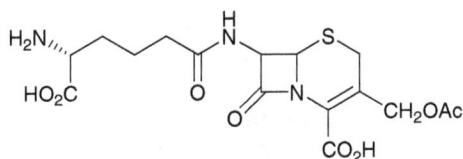

**1.23**

Several investigations of the fungus *Aspergillus fumigatus* between 1938 and 1945 by Raistrick and by Chain and Florey led to the isolation of compounds with high anti-bacterial activity and diverse structure. These included a quinone, fumigatin (**1.24**), a steroidal antibiotic, helvolic acid (**1.25**), and a diketopiperazine disulfide, gliotoxin (**1.26**). The latter had been isolated previously by Weindling in 1936 from *Gliocladium fimbriatum*. Griseofulvin was first isolated in 1939 by Raistrick from *Penicillium griseofulvum*. Studies of the antagonism between fungi led to the isolation in 1946 of griseofulvin from *P. janczewskii* as an anti-fungal metabolite with a particular 'curling' effect on the hyphae of the plant pathogen, *Botrytis allii*. The structure of griseofulvin (**1.27**), which has been used in medicine to treat fungal infections of man, was established in 1952. It is discussed in Chapter 4.

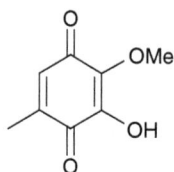

1.24

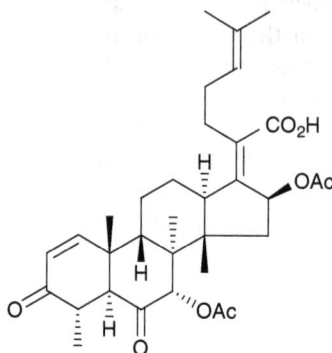

1.25

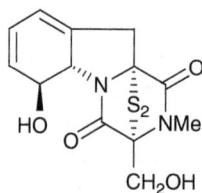

1.26

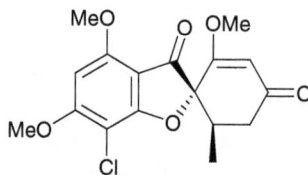

1.27

## 1.5.4 Study of Fungal Plant Diseases 1940–1965

The association of fungi with plant diseases has provided the stimulus for many studies. One particular study was that of the 'bakanae' stem elongation disease of rice. Examination of the causative organism, *Gibberella fujikuroi*, led to the isolation of a crude gibberellin in 1938. Careful microbiological work at ICI afforded a strain whose major metabolite was gibberellic acid. This was described in 1954 by Cross. The full structure and stereochemistry of gibberellic acid (**1.28**) was established in 1962 through a combination of chemical and spectroscopic studies by Cross and Grove and X-ray crystallographic studies by Scott. Work on the biosynthesis of gibberellic acid, which is discussed later, led to the isolation of many other gibberellins and compounds possessing an ent-kaurene skeleton. In 1958 gibberellins were identified as plant growth hormones in higher plants by MacMillan. The enzymology and genetics of gibberellin biosynthesis in *Gibberella fujikuroi* are aspects of continuing interest. The metabolites of some other phytopathogenic organisms, such as *Alternaria* species, were identified at this time (see Chapter 8).

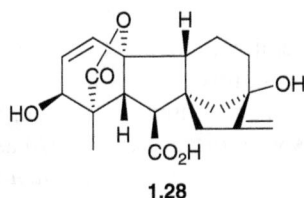

**1.28**

## 1.5.5 Impact of Spectroscopic Methods on Structure Elucidation

The development of modern spectroscopic methods from the 1950s onwards had an immense impact on the determination of the structures of fungal metabolites. The milligram scale on which these methods work meant that the amounts of material produced by different organisms in laboratory scale cultures, and that could be separated by chromatographic methods, could be studied. For example, correlations between the frequency of absorption and the structural environment of carbonyl groups played an important role in the determination of the structure of gladiolic acid (**1.29**) from *Penicillium gladiolii* in 1951 by Grove. The correlation tables published then have been used in many subsequent studies. The characteristic ultraviolet absorption of polyacetylenes was invaluable in their detection in the culture broths of many higher fungi in studies reported separately by Bohlmann and Jones between the 1950s and the 1970s. However, NMR spectroscopy has made the most important contributions. Some of the earliest applications of NMR spectroscopy to organic structure determination involved fungal metabolites such as gibberellic acid (**1.28**) (1959), cephalosporin C (**1.23**) (1961) and in correcting the structure of the trichothecenes (**1.30**) (1962). The structures of many mycotoxins and phytotoxins could not have been determined without the aid of $^1$H and $^{13}$C NMR spectroscopy. Advances in X-ray crystallography and in particular in the use of non-heavy atom methods have led to the solution of many natural product structures. The structure of a fungal metabolite, rosein III, was one of the first to be determined by direct methods in 1976.

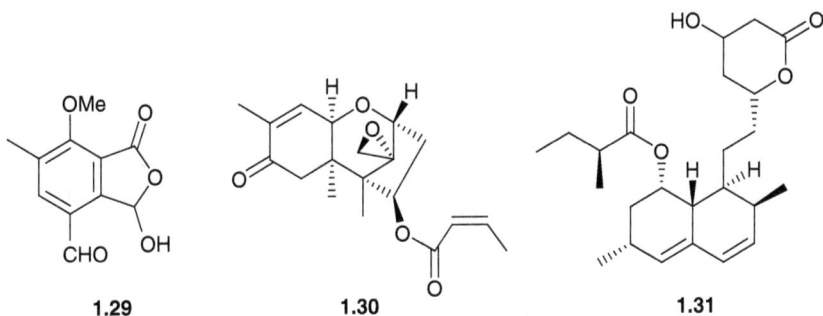

**1.29**     **1.30**     **1.31**

## 1.5.6 Fungal Metabolites 1965–2005

Over the last 40 years, many novel fungal metabolites have been isolated with quite diverse carbon skeleta. There have been two main driving forces behind this work. One has been the search for fungal metabolites with useful pharmaceutical activity; this has led to several compounds that have achieved commercial importance. These include the statins such as lovastatin (mevinolin) (**1.31**) from *Aspergillus terreus* (1974) which is used for the inhibition of cholesterol biosynthesis. The cyclosporins were discovered in 1970 as metabolites of the fungi *Cylindrocarpon lucidum* and *Tolypocladium inflation* and their undecapeptide structure was reported in 1976. They are widely used as immunosuppressant agents. The structure of aphidicolin (**1.32**) from *Cephalosporium aphidicola* was reported in 1972. This metabolite is a potent inhibitor of DNA polymerase α and has been examined as a potential tumour inhibitor and anti-viral agent. Pleuromutilin (**1.33**) was first described in 1951 from *Pleurotus mutilus* and its structure was established in 1962. It is the core for the antibiotic thiamulin®. The oudemansins (**1.34**) and strobilurins are a family of anti-fungal β-methoxyacrylates that were first isolated in 1977. They have subsequently been isolated from many Basidiomycetes, such as *Strobilurus tenacellus*. The β-methoxyacrylate moiety formed the lead structure for a novel range of commercial synthetic fungicides such as azoxystrobin®.

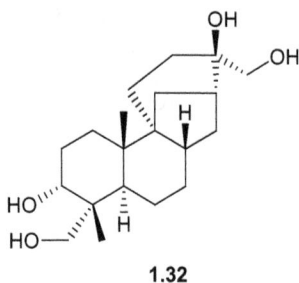

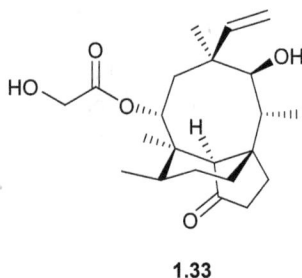

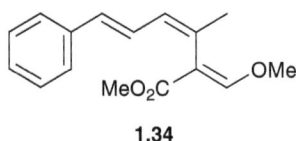

**1.32**                                                        **1.33**

**1.34**

The second major area of investigation has been into those fungi responsible for plant diseases. Particular attention has been paid to Dutch elm disease (*Ceratocystis ulmi*), to *Botrytis cinerea*, which is a pathogen on many commercial crops, and to those fungi that are pathogens on the Gramineae such as wheat, barley and rice. The phytotoxic metabolites of the honey-fungus, *Armillaria mellea*, have been thoroughly examined by several groups. The constituents of other Basidiomycetes, particularly those growing wild in Western Europe, and their biological activity has attracted attention especially through the work of Steglich. Many of these metabolites are sesquiterpenoids.

The microbiological hydroxylation of the steroid progesterone (**1.35**) at C-11 by *Rhizopus arrhizus* was reported in 1952. This biotransformation, which is now carried out commercially on a substantial scale, provided the facile means of introducing functionality at a site that was chemically difficult to access. It meant that many analogues of the cortical steroid hormones became available for the treatment of rheumatoid arthritis and other conditions. This observation has stimulated a great deal of work in the use of fungi for biotransformations, not just using steroidal substrates.

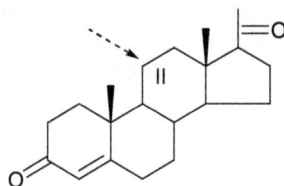

**1.35**

The association of fungi with mammalian diseases came to the fore in 1960 with the discovery of the aflatoxins (**1.36**). The death of turkeys from liver damage having been fed on groundnuts contaminated with *Aspergillus flavus* led to the isolation of the highly carcinogenic aflatoxins. These developments are discussed in Chapter 9. It led to the awareness of the potential human health hazards from microbial metabolites and the implications of the presence of other mycotoxins in foodstuffs such as patulin in apple juice and the trichothecenes on corn. The development of analytical methods for the detection of mycotoxins has become an important aspect of food science.

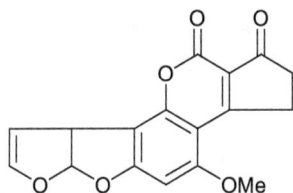

**1.36**

On a more positive note a mycoprotein known as Quorn®, which was obtained from *Fusarium venenatum* (originally described as a strain of *F. graminearum*), was approved for human consumption in the United Kingdom in 1984 and marketed as a meat substitute in the 1990s. The original fungal isolate was discovered in a soil sample from Marlow in the 1960s.

### 1.5.7 History of Biosynthetic Studies with Fungi

Microbial metabolites have proved to be more amenable to biosynthetic studies than the natural products obtained from higher plants partly because there are not the seasonal problems of production and also because the incorporation of labelled precursors is higher. In 1907 Collie had proposed that natural polyphenols might be biosynthesized from polycarbonyl compounds (polyketides) derived from linking acetate units together. This idea was taken up by Birch in 1953. The incorporation of carbon-14 labelled acetate by *Penicillium* species into 6-methylsalicylic acid (**1.37**) and griseofulvin (**1.27**) was described in 1955 and 1958, respectively. The discovery in 1958 that mevalonic acid was the precursor of the isoprene units of cholesterol in mammalian systems was followed by biosynthetic experiments on fungal metabolites, including gibberellic acid (**1.28**) (1958), rosenonolactone (**1.38**) (1958) and trichothecin (**1.30**) (1960). These metabolites were subsequently studied in greater detail in the 1960s and 1970s.

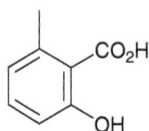

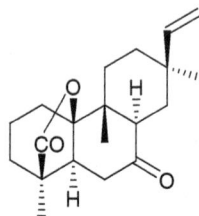

**1.37**                                          **1.38**

The radiolabelling experiments required careful degradation to establish the specificity of the labelling. Stable isotope studies using $^{13}$C NMR and more recently $^2$H NMR have overcome these problems and provided additional information. The first experiments using carbon-13 in the biosynthesis of fungal metabolites were described by Tanabe in 1966 on the biosynthesis of griseofulvin. The sites of the carbon-13 labels were located by measuring the $^{13}$C–$^1$H satellites in the $^1$H NMR spectrum. The first direct measurements of biosynthetic enrichments in the $^{13}$C NMR spectrum were made by Tanabe in 1970 in studies with radicinin (**1.39**) produced by *Stemphyllium radicinum*. The identification of $^{13}$C–$^{13}$C coupling patterns reported by Tanabe in 1973 in connection with a study of mollisin (**1.40**) biosynthesis provided an important method for establishing the integrity of units. The measurement of carbon-13 enrichments, coupling patterns and the use of carbon-13 as a 'reporter' nucleus for deuterium and oxygen-18 labelling studies, particularly in the work of Simpson, Staunton and Cane, have shed a great deal of light on biosynthetic pathways in fungi. Many of these studies are discussed in detail in later chapters.

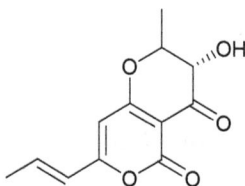

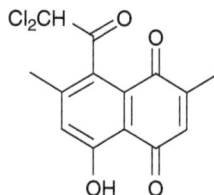

**1.39**                                          **1.40**

Microorganisms, particularly yeasts, had long been a source of enzyme preparations. Consequently, during the 1980s many enzyme systems associated with secondary metabolite biosynthesis were isolated. Once the acetate labelling patterns of polyketide metabolites were established in the 1970s the emphasis of biosynthetic studies moved towards the study of the polyketide synthases. The ideas built upon knowledge gained from the fatty acid synthases. Although studies on 6-methylsalicylic acid synthase and lovastatin polyketide synthase have been reported, the main thrust of the work since the late 1980s has been with erythromycin polyketide synthase obtained from a Streptomycete, *Saccharopolyspora erythraceae*, rather than with a fungus. The results of this work are discussed in the Chapter 4 on polyketides. An interesting facet of this work is that the genes that determine the structure of these enzymes are clustered together. The significance of this is discussed later. The chemical enzymology of some fungal terpene synthases, including some of the sesquiterpene and gibberellin systems, has also been examined with results that are discussed in the relevant chapters.

CHAPTER 2

# The Chemistry of Growing Fungi

## 2.1 The Culture Medium

The growth of fungi is dependent on the chemical composition of their environment. The difference between the autumnal microflora of an acidic heathland and that of chalk downland is quite a striking illustration. In this chapter we consider the role of the chemical composition of the medium on which a fungus grows first in the context of the laboratory and then in the 'wild' situation.

In the laboratory a fungus can be grown on 'surface' or 'still' culture, in 'shake' culture or in stirred and aerated fermentations. The extent of aeration differs between these and it affects not only the rate at which a fungus grows but it can also determine the metabolites that are produced.

Culture media for growing fungi can be either synthetic or natural. A typical synthetic medium will contain a carbon source, normally a sugar, a nitrogen source such as an ammonium salt, a phosphate, a magnesium and a potassium salt often also providing sulfate and chloride ions together with the salts of trace elements. Two common synthetic media are the Czapek–Dox and Raulin–Thom solutions. These differ primarily in their nitrogen source. The original medium designed by Raulin used ammonium nitrate as the source of nitrogen. As the ammonium ions were used up first, the medium became quite acidic. Thom substituted ammonium tartrate in place of the ammonium nitrate to give a medium with a pH nearer to 4. Glycine ($2\,\mathrm{g\,L^{-1}}$) is also sometimes added as a nitrogen source. The sugars may be either sucrose or glucose. Tables 2.1 and 2.2 give the composition of a typical medium. Metabolite production can be quite sensitive to the constituents of the medium. For example, it has been reported that the polyketide flavipin (**2.1**) was only obtained from *Aspergillus flavipes* when it was grown on a Raulin–Thom medium and not when it was grown on a Czapek–Dox medium. The medium on which Basidiomycetes grow may be more complex. A typical medium might include (per litre), glucose (50 g), peptone (20 g), yeast extract (2 g), potassium dihydrogen phosphate (0.87 g), magnesium sulfate (0.5 g), calcium chloride (0.5 g), corn steep liquor (5 mL) and trace metal solution (20 mL).

The Chemistry of Fungi
By James R. Hanson
© James R. Hanson, 2008

**Table 2.1** Some typical media.

| | Weight per litre distilled water (g) |
|---|---|
| **Czapek–Dox medium** | |
| Glucose | 50 |
| Sodium nitrate | 2 |
| Potassium dihydrogen phosphate | 1 |
| Potassium chloride | 0.5 |
| Magnesium sulfate | 0.5 |
| Iron(II) sulfate | 0.01 |
| **Raulin–Thom medium** | |
| Glucose | 50 |
| Tartaric acid | 3 |
| Ammonium tartrate | 3 |
| Ammonium hydrogen phosphate | 0.4 |
| Ammonium sulfate | 0.2 |
| Potassium carbonate | 0.4 |
| Magnesium carbonate | 0.3 |
| Iron(II) sulfate | 0.05 |
| Zinc sulfate | 0.05 |

**Table 2.2** Trace metal solution.

| Element | Weight per 100 mL distilled water (g) |
|---|---|
| Cobalt nitrate | 0.01 |
| Iron(II) sulfate | 0.1 |
| Copper sulfate | 0.015 |
| Zinc sulfate | 0.161 |
| Manganese sulfate | 0.01 |
| Ammonium molybdate | 0.01 |

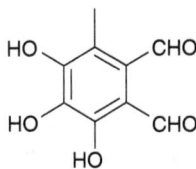

**2.1**

The medium may be sterilized by steaming at 100 °C several times over a few days or by autoclaving with steam at 120 °C under pressure for 15–20 min. With a complex medium it may be necessary to autoclave some of the components separately to avoid caramelizing the sugars.

The justification for these constituents can be seen in the structures of the primary metabolites of fungi. Many metabolic processes utilize phosphates, often with a magnesium ion. It has been estimated that a quarter to a third of all enzymes contain a metal ion as a functional component. Iron is particularly important in the cytochrome P450 oxidases which are involved in the biosynthesis of many fungal metabolites. Other metal ions, particularly zinc, play a significant role in stabilizing protein structures. Table 2.2 gives the composition of a typical trace metal solution. Excess metal ions can be toxic. Fungi can be particularly sensitive to copper ions – an active constituent of Bordeaux mixture. Some fungi will grow better if a small amount of a vitamin B supplement based on a yeast extract (*e.g.* Marmite®) is added.

Natural media are often based on corn steep liquor, malt extract or potato extract. Corn steep liquor is a by-product from the preparation of starch from maize and is particularly useful as a source of nitrogen. In the initial studies on the development of the penicillin fermentation the addition of corn steep liquor to the medium produced a significant increase in the antibiotic titre.

Unsurprisingly, since many fungi are plant pathogens, plant extracts can be useful constituents of media. Corn steep liquor or a solid medium of rice or grain can be helpful for fungi that affect aerial parts of a plant whilst a carrot or potato extract may be used with root-infecting fungi. Basidiomycetes may grow on wood chips.

The media that are used for growing fungi differ from those that are used for growing bacteria. Fungi require a slightly acidic medium (*ca.* pH 4) whereas bacteria grow better in a medium at a pH nearer neutrality or even under slightly alkaline conditions. Most fungi grow well between 18 and 25 °C whereas bacteria usually require a higher temperature (*ca.* 37 °C). The nitrogen source in bacterial media is often a protein hydrolysate as in bacteriological peptone. In general, bacteria grow much more rapidly than fungi. The effects of bacterial food poisoning are often apparent in hours whilst the irritation and skin lesions from the fungal infection known as athlete's foot (*Tinea pedis*) take several days to manifest themselves. Bacterial contamination of a fungal culture is often apparent quite early in the fermentation, when it may be detected by the smell and cloudy appearance of the culture.

## 2.2  Laboratory Fermentation

The sequence of operations in carrying out a laboratory fermentation involves first taking the master culture and preparing a series of sub-master slopes on a nutrient agar such as potato–dextrose agar. These are in plugged tubes or small bottles (slopes) in which the agar has been allowed to set at an angle exposing a large surface area. When these slopes have grown satisfactorily, a seed culture is prepared using a sterile medium in a plugged conical flask. When this has grown, usually for about 48 hours, aliquots are distributed through the production vessels. If the fermentation is to be grown on surface culture, a sterile shallow layer of the medium (*ca.* 125 cm$^3$) is inoculated in a flat one-litre Roux bottle. The offset neck of the Roux bottle is plugged with cotton wool to allow the fermentation to breathe whilst avoiding contamination from spores in the air. As many as 80–100 bottles may be inoculated. Throughout these transfers care has to be taken to avoid contamination by, for example, 'flameing' surfaces such as the wire loops and the neck of the flask, which may have come in contact with potential infection.

The early penicillin fermentations were grown in hospital bed pans and flasks based on this shape have been used quite widely. They can, however, be quite difficult to clean. Large conical flasks, whilst being quite bulky, can be more convenient, particularly if the fungus is being grown on a solid support such as rice, sterile wood chips, filter paper or glass wool impregnated with the liquid medium. Conical flasks are used for shake culture in which the fermentation is grown on an oscillating table at 100–120 rpm. The shaking movement increases the aeration and the dispersion of the mycelium. Sometimes, cylindrical vessels on rollers are used for cultures on solid media. In this case, rolling the vessel increases aeration. Stirred fermentations are carried out in pots that are stirred and aerated with sterile air. These fermentors are expensive but on an industrial scale they can operate with thousands of gallons of medium. Some of the early laboratory stirred fermentors were converted washing machines!

As a fermentation progresses, several parameters may be monitored. The three most common are the sugar level, measured by the optical rotation of an aliquot, the pH and the mycelial dry weight. In an industrial situation changes in other parameters such as oxygen levels and the availability of a nitrogen source may be measured during the fermentation.

Whilst a surface culture may be grown for a month or even longer, a shake culture may be grown for 10–14 days and a stirred fermentation for an even shorter period of time. Methods have been developed for continuous culture in which fresh sterile medium is added to a growing fermentation whilst spent medium containing the metabolites is removed. On the laboratory scale this is quite difficult to maintain sterile but it is used commercially.

Throughout a fermentation it is important to avoid contamination, particularly if several organisms are being grown in the same laboratory. It is unwise to attempt to grow bacteria and fast-growing yeasts in bulk in the same laboratory as slower-growing fungi. I remember a young student who was persuaded by the theoretical chemists (!) to brew some beer in the fermentation laboratory for a Christmas party using a rather vigorous yeast. The resultant cross-contamination had disastrous consequences for many subsequent slower-growing fungal fermentations. It is also unwise to bring collections from the wild into a growing room because these often harbour mites that can carry fungal spores with them on their travels round the laboratory. The relative constant temperature of a basement can provide a good growing area but it is worth remembering that mice and 'silver fish' also congregate in these areas. In addition, although a fermentation may be complete and the metabolites isolated, the spent fermentation medium can still provide nutrients for bacteria and fungi. Fermentation residues need to be treated with a disinfectant such as Stericol® or autoclaved and disposed of hygienically. Cleanliness in the fermentation laboratory is of paramount importance.

## 2.3 Isolation of Fungal Metabolites

There are various ways of harvesting a fermentation and isolating the metabolites. Some metabolites are excreted into the culture broth whilst others are found in the mycelium. The mycelium is filtered from the culture medium. Different metabolites may be found in the medium compared to the mycelium. The sesquiterpenoid metabolite of *Trichothecium roseum*, trichothecin (**2.2**), is found mainly in the culture filtrate whilst the diterpenoid rosenonolactone (**2.3**) is found in the mycelium of the same fungus. As a crude generalization, the extra-cellular metabolites isolated from the culture filtrate may be associated with the combative relationship of the organism with its environment whilst those isolated from the mycelium may have a protective role.

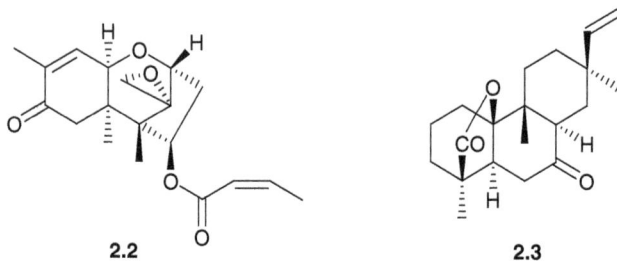

**2.2**                                                        **2.3**

Mycelial metabolites are best isolated by filtering the mycelium, drying it, preferably with gentle heat under vacuum, and then extracting the powdered mycelium in a Soxhlet thimble with a solvent such as chloroform. Metabolites from the broth are extracted with a solvent such as diethyl ether, ethyl acetate, butyl acetate or chloroform. There is, sometimes, a loose association between a metabolite and protein in the culture filtrate and the recovery of the metabolite may be improved by acidifying the broth to pH 2. However, care has to be taken not to generate artefacts arising from acid-catalysed reactions. An alternative method, particularly when large volumes are involved, is to add active charcoal (*ca.* $12\,g\,L^{-1}$) to the culture filtrate and then leave the mixture to stand in a cold-room overnight. The charcoal is then filtered off and the metabolites can be eluted from the charcoal with acetone. The metabolites are then separated through the standard natural product techniques by dividing them into acidic, neutral and basic fractions followed by chromatography. Typical yields of the major metabolites obtained from laboratory cultures, as opposed to industrial fermentations, are of the order of $50$–$100\,mg\,L^{-1}$ of the culture broth. In an exceptional case the yield may reach $1\,g\,L^{-1}$. However, in some cases the yield of an interesting metabolite may be as low as $1\,mg\,L^{-1}$.

It can happen that, after repeated sub-culturing, a fungus ceases to produce a secondary metabolite. This is not surprising given the role of some secondary metabolites in facilitating the growth of the fungus in a competitive environment. When this stimulus is removed, metabolite production ceases. Apart from returning to the original master culture, a solution to rejuvenating the organism can be to inoculate a host plant with the fungus and, when the plant shows signs of stress, re-isolate the fungus, possibly using a 'single spore' isolation technique to select a productive strain.

## 2.4 The Stages in a Fermentation

Several stages in the growth of an organism have been recognized as being determined by the chemical composition of the medium. The growth of the fungus *Gibberella fujikuroi*, which produces the plant hormone, gibberellic acid (**2.4**), has been studied in this context. The fungus was grown in stirred culture on media in which the initial glucose, nitrogen (ammonium nitrate), phosphate and magnesium ion concentrations were varied so that one of these would be exhausted before the others. Distinct growth phases were identified, including a balanced phase when all the nutrients were present. This was a period of rapid proliferation in which the mycelium maintained a constant form. When one of the nutrients was exhausted there was a transition to a maintenance phase in which internal resources, for example stored fat, were utilized before a terminal phase in which lysis of the mycelium occurred and some mycelial components were returned to the medium. If the concentration of sugar was too high, the fungus did not grow well. The initial concentration of the nitrogen source obviously determined the protein and DNA level in the mycelium. Gibberellic acid was not produced until the inorganic nitrogen source was exhausted from the medium. Thus there was a distinction between the growth and production phases.

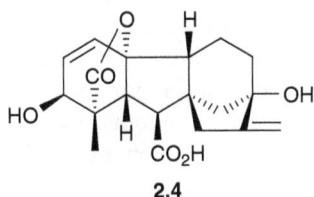

**2.4**

A more general picture of the relationship between the growth of a fungus and the production of metabolites is set out in a graph of mycelial dry weight against time (Figure 2.1). When the medium is first inoculated there may be a lag phase as the organism becomes established. This is followed by a rapid growth phase (tropophase = balanced growth phase). The organism then reaches the idiophase (maintenance phase) in which its characteristic metabolites are often produced. The occurrence of this stage may be influenced by the depletion or exhaustion of a particular nutrient. In some fermentations the lack of continued growth characteristic of this phase may be induced by the artificial addition of one of the metabolites of the fungus, *i.e.* there may be a feedback inhibition of growth. The idiophase is followed by a steady decline and lysis of the cells. The presence of cell constituents in the medium arising from lysis of the cells can sometimes make the recovery of metabolites difficult in the late stages of a fermentation. Some of the products of lysis may be sufficiently surface-active to cause significant emulsion problems on extraction with an organic solvent.

In studying fungal metabolites it is important to determine the stage in the fermentation at which the metabolites are produced, not just to obtain the maximum yield but also to establish the optimum time for biosynthetic studies.

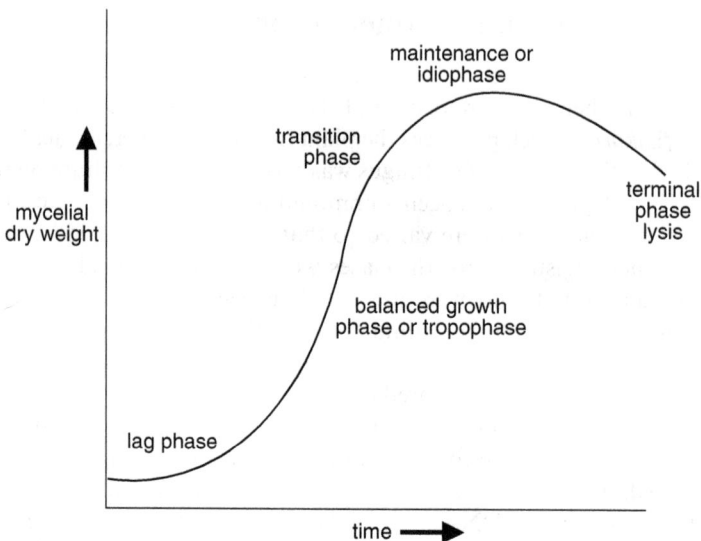

**Figure 2.1** Stages in a fermentation.

As noted previously, the formation of a particular metabolite, *e.g.* gibberellic acid (**2.4**) and the red pigment bikhaverin (**2.5**) by *Gibberella fujikuroi*, may be correlated with changes in fermentation parameters, in this case nitrogen levels in the medium, and these can affect the biosynthetic incorporation. Figure 2.2 shows the variation of the incorporation of [$^{14}$C]acetate with time into trichothecin (**2.2**) by *Trichothecium roseum* grown on shake culture.

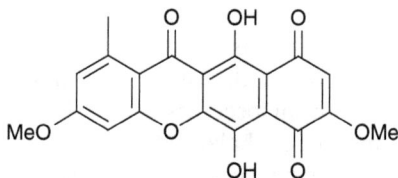

**2.5**

With a biotransformation it is also sensible to allow the fungus to become established, *e.g.* for two days after inoculation, before adding the substrate. The substrate for biotransformation may be added at a concentration of 100–500 mg per litre of medium in a solvent such as acetone, ethanol or dimethyl sulfoxide at a solvent concentration of 5–10 mL L$^{-1}$. The solution of the substrate should be added using a sterile pipette under sterile conditions and care should be taken to avoid the substrate precipitating onto the mycelium. The use of ethanol may encourage microbial reductions whilst dimethyl sulfoxide can be oxidized to the crystalline dimethyl sulfone. A biotransformation may take 5–10 days on shake culture and should be monitored by thin-layer chromatography. It is helpful to grow a small control fermentation including the solvent but without the added

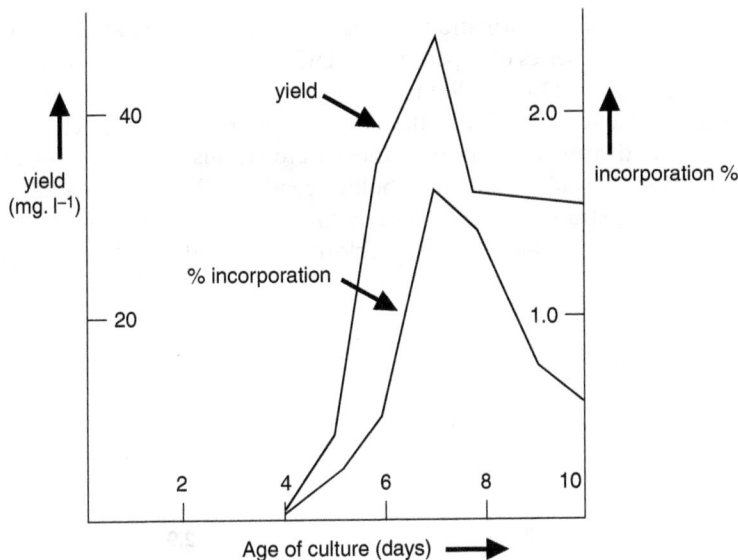

**Figure 2.2**   Incorporation of [1-$^{14}$C]acetate into trichothecin (**2.2**).

substrate. TLC of the metabolites of the control fermentation can help identify the novel products in the biotransformation. It is also worth remembering that these conditions can lead to aerial autoxidation. Some 'metabolites' may in fact be artefacts.

## 2.5   Utilization of the Constituents of the Medium

The measurement of glucose levels by polarimetry is a useful method of determining the carbon balance of fermentations as they progress. In some early studies on the isolation of fungal metabolites the depletion of glucose in the medium was taken as an indication of metabolite production. For example, *Penicillium griseofulvum* was grown on surface culture at 25 °C for 35 days, during which time the glucose concentration in the medium fell from 5% to 0.25%. The metabolite 6-methylsalicylic acid (**2.6**) was then isolated from the medium at the unusually high concentration of 1.1 g L$^{-1}$.

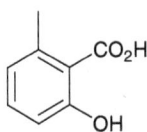

**2.6**

The depletion of other constituents of the medium can also be a guide to metabolite production. For example, the utilization of chloride ions by a fungus can be indication of the formation of chlorine-containing metabolites. In 1940 Raistrick reported a series of experiments in which 139 different organisms were grown on a Czapek–Dox medium containing 0.5 g KCl per litre. Whilst 57 organisms utilized less than 5% of the available chlorine, 77 used between 5 and 25% and 5 used over 25%. Two of these were strains of *Aspergillus terreus*, which produced the chlorinated metabolites geodin (**2.7**) and erdin (**2.8**), whilst another two were strains of *Caldariomyces fumago*, a 'sooty' greenhouse mould. This was shown to produce a novel dichloro-compound, caldariomycin (**2.9**).

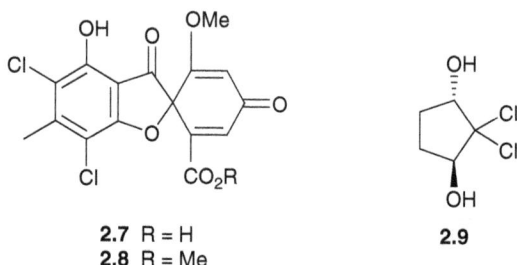

**2.7** R = H
**2.8** R = Me

**2.9**

Several fungi will produce volatile chloro compounds such as chloroform and methyl chloride. For example, *C. fumago*, when grown on surface culture produces about 70 μg L$^{-1}$ day$^{-1}$ of chloroform. In a study of 63 wood-rotting fungi, 34 different *Phellinus* species were shown to produce methyl chloride. This correlated with their ability to methylate compounds. If potassium bromide is substituted for potassium chloride, bromine can sometimes be introduced into the metabolites. This was used in the preparation of the bromo analogue (**2.10**) of the antibiotic griseofulvin (**2.11**). In the absence of halogen, dechlorogriseofulvin (**2.12**) was produced.

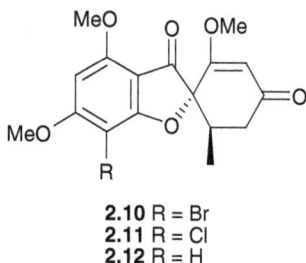

**2.10** R = Br
**2.11** R = Cl
**2.12** R = H

A similar study has been carried out on the utilization of sulfate. Amongst the organisms with a high sulfate consumption were *Penicillium notatum* and *P. chrysogenum*, which produce penicillin, and *Aspergillus fumigatus*, which produces the sulfur-containing metabolite gliotoxin (**2.13**). When the wood-rotting fungus *Schizophyllum commune* was grown on a medium containing ammonium sulfate it formed the unpleasant smelling methyl mercaptan. However, sulfites and sulfur dioxide are inimical to some fungi. Where sulfur dioxide is an aerial pollutant, black spot (*Diplocarpus rosae*) on roses is less common.

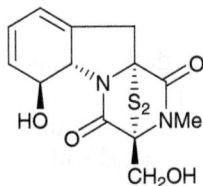

**2.13**

The ability of fungi to convert arsenites into the poisonous and volatile trimethylarsine has been associated with cases of arsenic poisoning. In 1839 Gmelin attributed the garlic odour in some damp rooms to poisonous volatile arsenic compounds that can be formed from mouldy wall paper. A green pigment, Paris Green or Schweinfurt Green, which is a double salt of copper acetate and copper arsenite, was used in wall papers in the 19th century. Several common organisms, including *Aspergillus*, *Mucor* and *Penicillium* species, were subsequently identified with this ability to metabolize arsenite. Studies by Gosic in the 1890s, who trapped the gases as their mercury(II) chloride complex, suggested that the gas was diethylarsine hydride, although this is now known to be unstable in air. However, later work by Challenger in 1933 using *Penicillium* (*Scopulariopsis*) *brevicaulis* growing on bread impregnated with arsenious oxide showed that the gas was trimethylarsine. As late as 1932, fatal poisonings in this country were attributed to this microbiological action on wall paper, and it is alleged to have affected Napoleon in his imprisonment on St. Kitts. Notably, a common wood-rotting fungus, *Lentinus lepideus*, can metabolize the wood preservative, zinc arsenite this way. Selenites are also converted into dimethyl selenide by *Scopulariopsis brevicaulis*.

The ability of fungi to methylate constituents of their growth medium and produce volatile compounds can contribute to the spoilage of foodstuffs. 2,4,6-Trichlorophenol (**2.14**) is used in cardboard as a preservative whilst the tribromo analogue is used in some corrugated board as a flame-retardant. In a survey of over 100 fungi found as spoilage organisms on cardboard packaging, 17 were shown to convert 2,4,6-trichlorophenol (**2.14**) into 2,4,6-trichloroanisole (**2.15**). The ubiquitous fungus *Paecilomyces variotii*, can convert 2,4,6-tribromophenol into 2,4,6-tribromoanisole. These volatile methyl ethers, which have a musty odour, have been detected in foodstuffs such as sultanas and cocoa that had been stored in this type of packaging which has become damp and infected by this fungus during storage. Cork that has been attacked by organisms such as *Armillaria mellea* accumulates volatile metabolites arising from its degradation, which can then contaminate wine and contribute to its 'corked' odour.

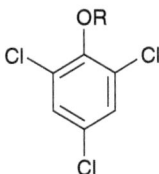

**2.14** R = H
**2.15** R = Me

## 2.6 Fungi Growing in the Wild

The chemistry of fungi growing in the wild is affected by their environment. Many Basidiomycetes grow best on sandy woodland soil. They are often mycorrhizal, drawing their nutrients in association with the roots of particular plants. Thus, particular organisms are often associated with particular trees. Some orchids will not grow without their associated fungi. A warm moist summer favours the growth of the underground mycelium which in the autumn triggers the formation of fruiting bodies. The mycelium may persist for many years and cover a substantial underground area compared to the size of the fruiting body.

Just as metal ions affect the growth of fungi in liquid culture so they do in the wild. The spread and persistence of the underground mycelium leads to the concentration of metal ions. In mainland Europe the collection of edible wild mushrooms is much more common than in the UK. In the Czech Republic it has been suggested that in some areas as many as two-thirds of families collect wild mushrooms, with an average annual consumption of 7 kg. However, fungi collected near the sites of former smelters, landfill sites and land treated with sewage sludge can accumulate significant quantities of metal ions such as those of cadmium, mercury, lead, copper and chromium. Some organisms have a concentration factor of 50–300-fold for cadmium and 30–500-fold for mercury. Cadmium at levels as high as 100–300 mg $kg^{-1}$ dry weight have been reported in *Agaricus* species. When petrol containing lead tetraethyl anti-knock was in use, quite high levels of lead were detected in an *Agaricus* species collected adjacent to a main road in North London.

The concentration of hazardous material was highlighted as a result of the Chernobyl accident in 1986. In the following year $^{137}$cesium was detected in the fruiting body of the bay bolete, *Xerocomus badius* growing near Bonn approximately a thousand miles from Chernobyl. The organism concentrated the $^{137}$cesium in its pigments, where it replaced potassium in complexes with an appropriately named pulvinic acid pigment, norbadione (**2.16**). Measurable concentrations of contamination by $^{137}$Cs have been found in 98 species of mushrooms growing wild in the Ukraine and further afield.

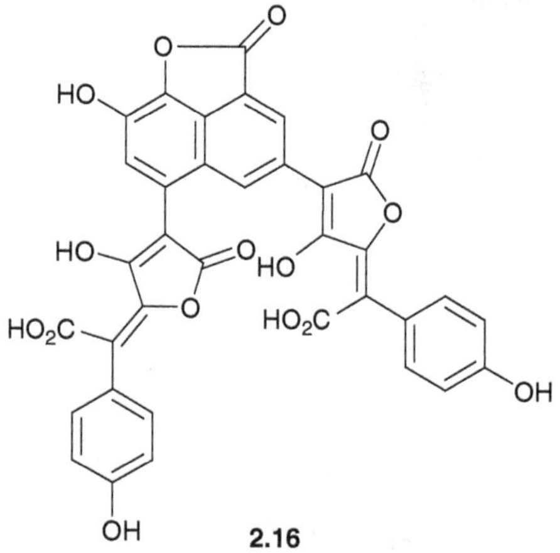


**2.16**

# 2.7 Biosynthetic Experiments

At first sight the diversity of the structures of fungal metabolites may seem to be immense. However, underlying the assembly of these structures is a series of common biosynthetic building blocks and pathways that provide a unifying feature. Thus it is possible to see various amino acids, acetate and isoprene units, together with extra carbon atoms arising from the methyl group of methionine, as components of a structure. These building blocks and pathways are common to the natural products found in higher plants. Nevertheless, although the assembly of fungal metabolites utilizes these common building blocks, there are some significant differences from compounds that are found in higher plants. These will become apparent in the following chapters.

The aim of biosynthetic experiments with fungal metabolites is to establish the structure of the building blocks, the order in which they are assembled, the way in which chains are folded to form the carbon skeleton and the sequence inter-relating precursors with the final metabolite. Biosynthesis is concerned with both sequences and reaction mechanisms. The sequence of the biosynthetic events, the role of intermediates and the stereochemistry of enzymatic reactions can be studied with appropriately isotopically-labelled substrates and with structural analogues of the natural intermediates. The chemical enzymology of individual steps and the role of key components and structures of the enzyme may be studied with isolated enzyme systems obtained from fungi. The features that determine the function of the enzyme and which control its activity may be determined by genetic studies in which mutants play an important role.

Biosynthetic experiments may be carried out with intact organisms, with crude cell-free systems or with purified enzyme systems. The enzymes within a fungal cell are highly organized. If this organization is disrupted by destroying the cell wall and subsequently removing the cell debris by centrifugation, a cell-free system can be obtained. This may be able to carry out just part of a biosynthetic sequence, enabling these steps to be studied in more detail. Genetic methods have enabled important enzyme systems to be overexpressed in mutant organisms or transferred to a bacterium such as *Escherichia coli*, from which the purified enzyme can be isolated in quantity.

Fungal metabolites are rarely produced alone. Preliminary evidence of biosynthetic sequences can be obtained by an examination of the structures of co-metabolites, their time of formation and by growing the organism under different conditions. The early years of biosynthetic experimentation were dominated by the use of radio-isotopic labelling techniques, particularly with tritium and carbon-14. Although today many of these experimental methods have been superseded by the use of stable isotopes, they are still of value in determining the optimum time for feeding experiments, for establishing some biosynthetic sequences and for examining some aspects of stereochemistry. The measurement of the changes in $^3H : {}^{14}C$ ratios between doubly-labelled substrates and metabolites can be particularly useful in establishing the fate of hydrogen atoms.

An objection to a biosynthetic result, particularly if the incorporation of the putative precursor is low, is that the possible precursor is being degraded to a component of the general metabolic pool and then incorporated. With radio-isotopic labels it was necessary to degrade the metabolite to demonstrate specificity of labelling. However, with stable isotopes, spectroscopic (particularly NMR) methods are sufficiently site-specific to obviate this. The choice of the site in a molecule for a label is obviously of paramount importance. Another objection, particularly with a putative intermediate that is late in a biosynthetic sequence, is that the compound being tested is not a genuine 'natural' intermediate but is being converted into the target molecule by an 'un-natural' induced process. The test is that the putative precursor must be formed and metabolized at a rate consistent with its intermediacy. Its formation can be established by trapping experiments in which a labelled earlier precursor is added to the fermentation together with the putative intermediate. This possible intermediate is re-isolated after a short time and examined to see if it has trapped any of the label.

The stable isotopes, carbon-13, deuterium and oxygen-18, together with the use of NMR and mass spectroscopic analyses, have played important roles in biosynthesis. The natural abundance of $^{13}C$ is 1.1% and the $^{13}C$ NMR signals of most fungal metabolites can be assigned. Simple substrates such as sodium acetate can be obtained with very high enrichments of $^{13}C$ (99%) at individual sites and, consequently, site-specific incorporations can be determined by examining the enhancement of the $^{13}C$ NMR signals in the spectrum of the metabolite. The use of $^{13}C_2$ doubly-labelled substrates enables $^{13}C-^{13}C$ coupling patterns to be observed in the metabolite and hence the integrity of bonds throughout a biosynthesis can be established. In carrying out this type of experiment with precursors such as [1,2-$^{13}C_2$]acetate, the labelled acetate is diluted with unlabelled material so that metabolites are not biosynthesized containing two adjacent labelled acetate units, thus avoiding a complicated labelling pattern in the metabolite. On the other hand, an informative $^{13}C-^{13}C$ coupling may be generated in a metabolite from the biosynthetic rearrangement of a highly-enriched multiply-labelled precursor.

Deuterium and even $^{18}O$ can induce a small isotope shift in the position of a $^{13}C$ signal and thus $^{13}C$ can be used as a 'reporter' nucleus for the presence of these isotopes. The $^2H$ NMR spectrum and the presence of $^2H-^{13}C$ couplings have also been used. There have been a few experiments using $^{17}O$ NMR. However, $^{18}O$ is more commonly detected by mass spectrometry.

Subsequent chapters give many examples of the perceptive and intellectually stimulating use of isotopic labels to discriminate between biosynthetic pathways.

Despite their apparent structural diversity, fungal metabolites can be grouped together in terms of their biosynthesis. The first group are those that are derived from amino acids. In these metabolites the amino acid is often incorporated intact, utilizing peptide-like bonds. The amino acids are not restricted to the proteogenic amino acids and some belong to the D- rather than the L-series. A second large group are formed by the fatty acid : polyketide synthase pathway primarily from acetate units. These metabolites can be further subdivided in terms of the number of acetate units that are involved in their formation. A third large group are the terpenoids, derived by the oligomerization of the $C_5$ unit isopentenyl diphosphate. This can be formed from acetate units *via* mevalonic acid or from 1-deoxyxyulose. Whereas in plants and bacteria both pathways are operative, the mevalonate pathway predominates in fungi. A fourth group of fungal secondary metabolites are formed from intermediates in the citric acid cycle. Whereas the shikimic acid pathway to form $C_6$–$C_3$ compounds is of major importance in the formation of secondary metabolites plants, it is of less importance in this context in fungi. A common feature of fungal metabolites is the formation of compounds that are derived by a combination of pathways.

Unlike primary metabolites, the genes that regulate the formation of the enzymes of secondary metabolite biosynthesis are often clustered. In several cases the loci of these genes have been determined. This has considerable significance in the control of secondary metabolite biosynthesis. The genes that code for several important polyketide pathways such as those leading to the aflatoxins and the statins have been identified. Similar work has also been reported for penicillin biosynthesis and some non-ribosomal peptides as well as terpenoid pathways such as that leading to the gibberellins.

The genes that determine secondary metabolite biosynthesis are not always constitutively expressed and often require transcriptional factors for their expression. Some genes may be silent or become silent on repeated sub-culturing as these factors are lost. Some pathways require specific proteins or small molecules to promote the expression of biosynthetic genes. This provides the rationale for re-invigorating a phytopathogenic organism by growing it on its host plant. The expression or suppression of other pathways may be dependent on more general factors such as pH, carbon or nitrogen source. For example, the gibberellin cluster in the fungus *Gibberella fujikuroi* is repressed by high nitrogen levels such as ammonium ions and glutamine. The regulation of genetic information leading to secondary metabolite biosynthesis plays an important role in fungal chemical biology.

CHAPTER 3
# Fungal Metabolites Derived from Amino Acids

## 3.1   Introduction

Amino acids are common building blocks in nature that are used not only for the macromolecules involved in cellular structures and in enzymes but also to form smaller secondary metabolites. Fungi, like other living systems, use amino acids as biosynthetic building blocks. However, there are several significant differences compared to higher plants in the way in which fungi metabolize amino acids in the formation of their secondary metabolites.

The common amino acids used in mammalian protein synthesis belong to the L-enantiomeric series. However, fungi also employ the D-enantiomers in the biosynthesis of some secondary metabolites. These are normally formed from the corresponding L-amino acid. Fungi can also make amino acids with structures that differ from those commonly found in mammalian proteins and in higher plants. These unusual amino acids are utilized for the synthesis of secondary metabolites and some peptides.

Many of the modifications of the amino acid units that are found in fungal metabolites differ from those in higher plants. In particular, the decarboxylation of amino acids is far less common in fungi. For example, the decarboxylation of phenylalanine or tyrosine to form an arylethylamine is far less common in fungi and hence alkaloids derived from β-arylethylamine units such as the benzyliso-quinoline alkaloids, which are common in plants, are unusual in fungi.

In this chapter we consider some important fungal metabolites that are biosynthesized by fungi such as *Penicillium* and *Aspergillus* species. These include the penicillin and cephalosporin β-lactam antibiotics. Some fungal metabolites derived from amino acids that have achieved notoriety because of their toxicity are described in Chapter 9 on mycotoxins. These include several tryptophan derivatives. Other compounds also derived from amino acids are to be found amongst the fungal pigments described in Chapter 7.

---

The Chemistry of Fungi
By James R. Hanson

# 3.2 Penicillins

The history of the discovery of the penicillins has been described in Chapter 1. Many of the developments in fermentation methodology have their origin in the production of the penicillins. The organism that was originally used for the production of penicillin in Oxford in the early 1940s was *Penicillium notatum*. However, the collaborative studies with American workers at the Northern Regional Research Laboratories in Peoria, Illinois, led to the identification of a more efficient strain of *Penicillium chrysogenum*, a mutant of which, Wis49–133, had yields of penicillin that were many hundred times that of the original source. Penicillin production was also found to be a function of the constituents of the medium and the conditions of growth. One of the big changes that the American workers made to the fermentation was to introduce corn steep liquor into the medium and to grow the fungus in stirred aerated tanks. This had the effect of changing the penicillin that was produced. It was also realized that that the penicillins were the amides of a common core with different side-chains (R in **3.1**). Whereas the British workers obtained mostly penicillin F in which the side-chain was the $\Delta^2$-pentenyl moiety, the American workers obtained penicillin G in which the side-chain was a benzyl group. Small amounts of *p*-hydroxybenzylpenicillin (penicillin X) and n-heptylpenicillin (penicillin K) together with n-amylpenicillin were also detected. In penicillin V, R is the phenoxymethylene ($PhOCH_2$) group. n-Amylpenicillin was known as gigantic acid and was obtained from *Aspergillus giganteus*.

The structure of the penicillins (**3.1**) was established in the days before spectroscopic methods were routinely applied to structure determination. The degradative evidence for the structure of the penicillins is best understood if it is realized that the central carbon (C-5) of the thiazolidine ring in the core of the structure is a masked aldehyde and, secondly, that the β-lactam is a strained four-membered ring in which the lactam does not behave as a typical amide. The shape of the ring precludes the normal amide resonance and hence hydrolysis of the lactam takes place more easily than would be expected for an amide. This hydrolysis then places a carboxylic acid in the β-position to the masked aldehyde so that decarboxylation of a β-keto acid can occur.

The strategy for the structure elucidation involved degradation by several contrasting and structurally revealing sequences, which afforded simpler fragments that could be synthesized by unambiguous methods (see Scheme 3.1). The same fragments were obtained by different degradations.

Two fragments were isolated by hydrolysis of the penicillins with acid. The first, penicillamine (**3.2**), was identified as β,β-dimethylcysteine by synthesis and shown to possess the unusual D-amino acid configuration. The second fragment, which retained the side-chain, was known as a penaldic acid (**3.4**). It behaved as a β-keto acid and lost carbon dioxide to give a penilloaldehyde (**3.5**). The penaldic acids were obtained by alkaline hydrolysis of a penicillin. Initially, this gave a dicarboxylic acid known as a penicilloic acid (**3.3**). Further

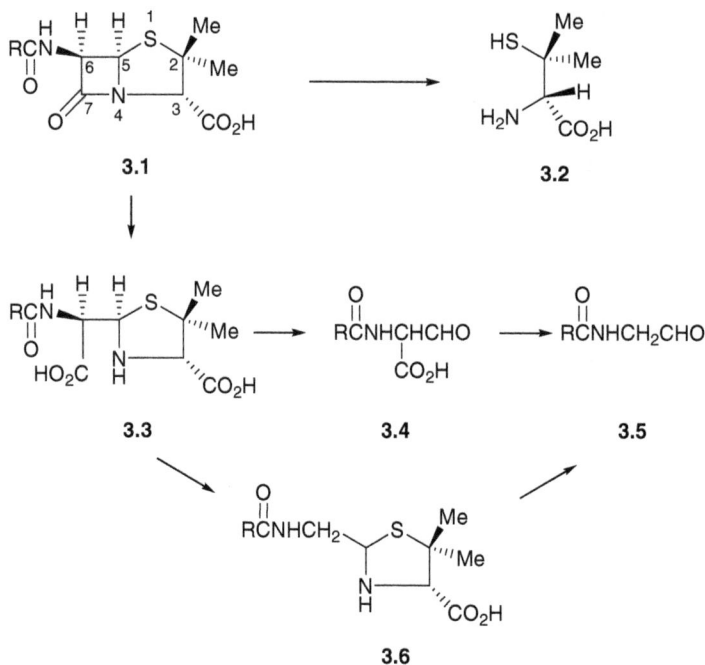

**Scheme 3.1**   Key degradations of penicillin.

hydrolysis led to the loss of the penicillamine unit and the formation of the penaldic acid. The penilloaldehydes were also obtained by treatment of the penicilloic acid with acid. This led to the evolution of carbon dioxide and the formation of a monocarboxylic acid, a penilloic acid (**3.6**). Removal of the sulfur from the penilloic acid with mercury(II) chloride and hydrolysis of the product gave the penilloaldehydes.

All of this evidence pointed to the presence in the penicillin of an aldehyde masked by the penicillamine and also to a group which when hydrolysed placed a carboxylic acid in the β position to the masked aldehyde. The conversion of penicillin into the penicilloic acid containing this carboxyl group involved the addition of a molecule of water. It also led to the release of a basic amino group. Although in retrospect it may now seem surprising, some time elapsed before this was seen as the hydrolysis of a cyclic four-membered ring amide. A distraction was provided by a series of compounds that arose by rotation about the C-5–C-6 bond of the penicilloic acid and condensation of the basic amino group with the side-chain amide. These penillic acids (**3.7**) were formed by an acid-catalysed hydrolysis of the β-lactam ring and cyclization to form an imidazole ring. Decarboxylation led to the formation of the penillamines (**3.8**). A wealth of heterocyclic chemistry was uncovered during the structural studies on the penicillins. Eventually, a β-lactam structure for the penicillins was accepted on the basis of an X-ray crystal structure described in 1945.

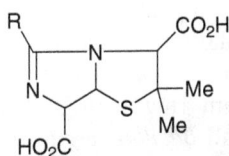

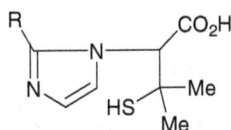

**3.7**          **3.8**

The instability of the penicillins owes much to the strained nature of the β-lactam ring in which the carbonyl group does not behave like that of an amide carbonyl and is more reactive to nucleophiles. The nitrogen is more basic than that of a typical amide. Hydrolysis of the β-lactam takes place more rapidly than would be anticipated for a typical amide. Neighbouring group participation from the carbonyl oxygen of an amide at C-6 facilitates this process. Enzymatically this process is catalysed by β-lactamases. These enzymes are produced by many bacteria that are resistant to penicillins. A metabolite of an Actinomycete, *Streptomyces clavuligenus*, clavulanic acid (**3.9**) has found considerable use as a β-lactamase inhibitor. It is a β-lactam itself and it is hydrolysed by the β-lactamase. However, the product reacts with the enzyme, thereby deactivating it. Thus, clavulanic acid is a suicide inhibitor. When it is administered together with a penicillin such as amoxycillin [R=4-HOC$_6$H$_4$CH(NH$_2$) in **3.1**] in a preparation known as augmentin$^{®}$, some of the bacterial resistance arising from the β-lactamase is overcome by the clavulanic acid and the amoxycillin continues to function.

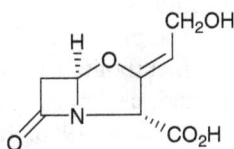

**3.9**

Implicit in the isolation of penicillins with different side-chains is the formation of a common core. Indeed, different penicillins, such as penicillin V could be produced biosynthetically by adding different side-chain precursors, *e.g.* phenoxyacetic acid, to the fermentation medium. The common core, 6-aminopenicillanic acid (H in place of R.CO in **3.1**), was described as a fermentation constituent in 1959. In 1946 a bio-autographic method of assaying the fermentation constituents was introduced. The penicillins were separated by paper chromatography and the antibiotic zones were located by incubating the paper in contact with agar seeded with *Bacillus subtilis*. This method was utilized in the detection of 6-APA. Repetition of some unexpected results of Kato obtained in 1953 from fermentations to which no side-chain precursors had been added led, in the late 1950s, to the isolation of 6-aminopenicillanic acid. When a poorly bio-active spot was sprayed with the side-chain precursor, phenylacetyl chloride, a bioactive zone developed. The spot was isolated and shown to be 6-aminopenicillanic acid.

Since the fermentation methods for producing modified penicillins were limited by the suitability of substrates, the isolation of this penicillin core became a major objective. Although many of the resistant strains of bacteria deactivate penicillins by opening the β-lactam ring with a β-lactamase, a number, including strains of *Escherichia coli* and *Bacillus cereus*, can cleave the side-chain with a penicillin acylase. Preparation of this enzyme on a large scale and its utilization with the natural penicillins have allowed 6-aminopenicillanic acid to be prepared commercially. Semi-synthetic penicillins such as ampicillin and amoxycillin can then be made by acylating the amino group of the 6-aminopenicillanic acid with the relevant acid chloride. The annual production of 6-aminopenicillanic acid has been estimated to be around 7000 tonnes.

# 3.3   Cephalosporins

The valuable anti-bacterial activity of the penicillins and the relatively easy bio-assay stimulated the search for other microbial metabolites with anti-biotic activity. Several useful substances, such as the tetracyclines, the macro-lide antibiotics and chloramphenicol, were discovered, particularly from the Actinomycetes. However, the cephalosporins were obtained from the fungus *Cephalosporium acremonium*. The particular active strain, known as the Brotzu strain, was discovered in the microbially highly competitive environment of a sewage outfall in Sardinia in 1945 by Guiseppe Brotzu. This species produced several different antibiotics. In 1954 one of these, which was originally known as cephalosporin N (now penicillin N), was shown to be D-δ-aminoadipoyl-6-aminopenicillanic acid (**3.10**). The presence of the penicillin nucleus was established by degradation of this metabolite to products that were common to the penicillins. A second compound, cephalosporin C, was separated from the fermentation products and described in 1956. Its structure (**3.11**) was established in 1961 by a combination of chemical degradation and spectroscopic methods, including $^1$H NMR spectroscopy. Comparison between the elucidation of this structure and that of penicillin reveals the growing impact of spectroscopic methods on natural product chemistry in the 1950s and 1960s. In this context it is worth noting that the evidence for the structure of the β-lactam, clavulanic acid, obtained in the 1970s, was almost entirely spectroscopic. Although there was a close similarity between cephalosporin C and cephalosporin (penicillin) N, there were some striking differences in the spectroscopic properties. The presence of the β-lactam in cephalosporin C was revealed by degradation to δ-aminoadipoylglycine, which was also obtained from cephalosporin N. The β-lactam had a characteristic infrared absorption. The ultraviolet spectrum revealed the presence of an α,β-unsaturated carbonyl group. The $^1$H NMR spectrum showed that the structure differed from that of the penicillins by lacking the signals associated with the gem-dimethyl group and containing, instead, the resonance associated with an acetoxyl group. The key to the presence of the dihydrothiazine bearing an unsaturated acid came from the formation of an α,β-unsaturated lactone (cephalosporin C$_\alpha$)

on hydrolysis of the acetate. Hydrogenolysis of the sulfur with Raney nickel and hydrolysis of the resultant enamine gave the enol of β-methyl-α-oxo-γ-butyrolactone (**3.12**) which was identified by synthesis. The structure and stereochemistry of cephalosporin C was finally established by X-ray crystallography.

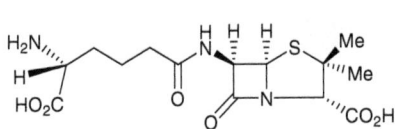

**3.10**

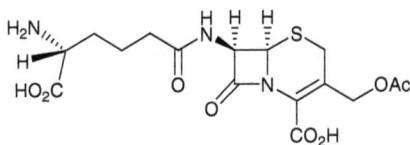

**3.11**

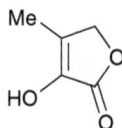

**3.12**

Like the penicillins, the side-chain of the cephalosporins has been cleaved from the cepham core by cephalosporin acylases and a series of semi-synthetic cephalosporins have been prepared.

# 3.4 Biosynthesis of β-Lactams

The biosynthesis of the penicillins and the cephalosporins has been thoroughly investigated (Scheme 3.2). Early studies had established that the side-chain of the final penicillin, as in penicillins F, G and V, depended on the constituents of the medium and that a tripeptide, δ-(L-α-aminoadipoyl)-L-cysteinyl-D-valine (LLD-ACV, sometimes known as the Arnstein tripeptide) (**3.13**), was shown to be the key intermediate in the formation of the core of the penicillins and cephalosporins. The first penicillin to be formed was isopenicillin N. By 1980 the crude enzyme system, isopenicillin N synthase, had been isolated. It required iron, dioxygen, ascorbate and a thiol in order to function. Stereospecific labelling studies revealed that the formation of the new C–N and C–S bonds in the ring system proceeded with retention of configuration. The oxidation of the thiol of cysteine was a stereospecific process with particular hydrogen atoms being removed. However, it was difficult to detect intermediates by direct methods. Nevertheless, some conclusions were drawn from kinetic isotope effects using specifically deuteriated amino acid precursors. This showed that the first C–H bond to be broken was that at C–3 of the cysteine residue. This evidence also pointed to a stepwise rather than a concerted process in which an intermediate is formed that remained enzyme bound before undergoing cyclization to form isopenicillin N.

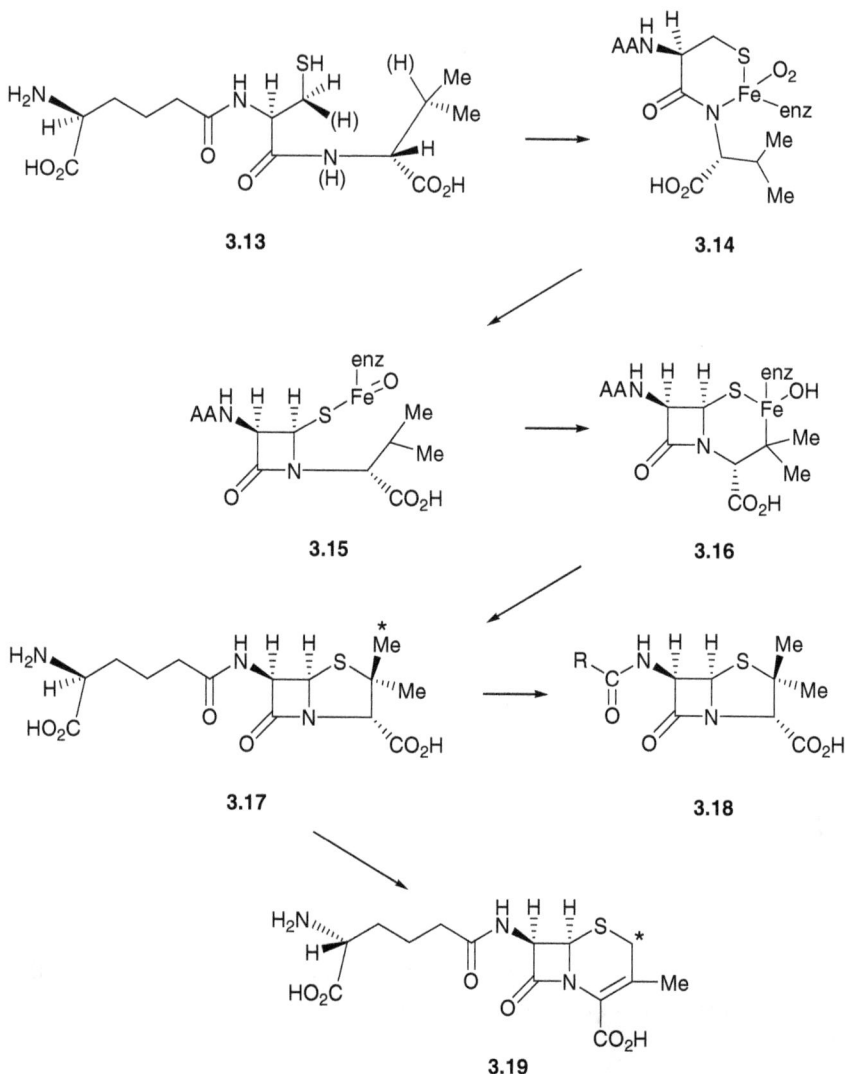

**Scheme 3.2**   Biosynthesis of the penicillins and cephalosporins.

The enzyme system, isopenicillin N synthase (IPNS), was cloned into the bacterium *Escherichia coli* to make more material available. This meant that the chemical enzymology of this biosynthesis could be studied more thoroughly. Various amino acids were then used in place of valine. When a free radical is formed adjacent to a cyclopropane ring it undergoes a ring-opening reaction. The ease of ring-opening of analogues of valine containing a cyclopropane ring adjacent to the carbon atom to which the sulfur becomes attached suggested that free radical intermediates were involved in this step in the biosynthesis. When allyl probes were used some of the metabolites picked up an oxygen substituent, suggesting that a feryl iron enzyme intermediate might be involved. The cyclization sequence is set out in **3.13–3.18**. Central to an understanding of this biosynthesis is the ease with which sulfur forms bonds to iron and undergoes one-electron oxidation.

The IPNS enzyme system has been examined thoroughly and shown to contain 336 amino acid residues. It has been used to synthesize a range of novel β-lactams, some of which show biological activity comparable to that of the natural penicillins. X-Ray crystallographic studies of isopenicillin N synthase in the presence of iron and ACV analogues have been carried out. By using nitric oxide as a structural mimic of molecular oxygen, a picture of the active site has emerged in which the oxygen is held close to the relevant sites that are oxidized.

Ring expansion of the penicillins to the cephalosporins has been examined. Following the elucidation of the structure of penicillin N and cephalosporin C, small amounts of isopenicillin N with an L-aminoadipoyl side-chain and de-acetoxycephalosporin C were found in *Cephalosporium acremonium*. Cell-free systems were obtained from the fungus which catalysed the ring expansion of penicillin N to deacetoxycephalosporin C (**3.19**) and the hydroxylation of this. A radical mechanism has been proposed in which the 3(pro-*R*)-methyl group of valine (marked * in **3.17**) becomes the ring methylene (marked * in **3.19**) of cephalosporin C.

## 3.5 Metabolites Containing a Diketopiperazine Ring

Just as the penicillins and cephalosporins can be viewed as modified peptides so can the diketopiperazines. The diketopiperazine ring and various modifications are found in several fungal metabolites derived from two amino acids. Simple diketopiperazines such as the anhydride **3.20** of di-L-phenylalanine have been isolated from *Penicillium nigricans*. Echinulin (**3.21**), from *Aspergillus echinu-latus* and *A. amstelodami*, has been shown to be derived from L-alanine and L-tryptophan. The three isoprene units are formed from mevalonate. Flavacol (**3.22**), which is a metabolite of *Aspergillus flavus*, is derived from L-leucine anhydride. Various cyclic hydroxamic acid derived from these piperazines have also been obtained from *Aspergillus* species. Aspergillic acid (**3.23**) and hydroxyaspergillic acid (**3.24**) are also biosynthesized from leucine and iso-leucine by *A. flavus*. Aspergillic acid has quite powerful antibiotic properties but it is too toxic to be used by man. Typical of a hydroxamic acid, aspergillic acid forms metal complexes. The complex with iron(III) is a red pigment, ferriaspergillin, which has also been isolated from *A. flavus*.

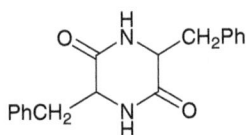

**3.20**

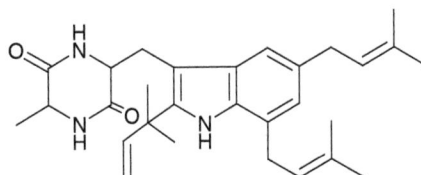

**3.21**

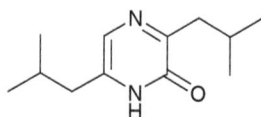

**3.22**

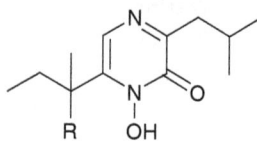

**3.23** R = H
**3.24** R = OH

### 3.5.1 Mycelianamide

The metabolites of *Penicillium griseofulvum* have attracted a lot of interest. Extraction of the mycelium furnished a polyketide griseofulvin (see Chapter 4) and an amide, mycelianamide (**3.25**). The latter was first reported in 1931 by Raistrick and a method for its separation from griseofulvin was described in 1939. Further evidence for its structure was presented in 1948 and the complete structure of mycelianamide was established in 1958 by Birch. This metabolite gave an intense reddish-brown colour with ferric chloride. Although it contained two nitrogen atoms, it had no basic properties but behaved as a very weak acid. Hydrolysis provided some useful identifiable fragments. Treatment with concentrated hydrochloric acid gave ω-amino-*p*-hydroxyacetophenone whilst dilute sulfuric acid gave a hydrocarbon, $C_{10}H_{16}$ (mycelene). Hydrolysis with dilute sodium hydroxide gave an optically inactive monobasic acid from which mycelene and *p*-hydroxybenzoic acid were obtained by acid hydrolysis. Mycelene was shown to be the hydrocarbon **3.26**. Reduction of mycelianamide with zinc in acetic acid removed two oxygen atoms and gave deoxymycelianamide, suggesting that mycelianamide was an N-oxide. Unlike mycelianamide this reduction product was insoluble in alkali and gave no ferric chloride colour. Hydrolysis of deoxymycelianamide gave alanine and *p*-hydrophenylpyruvic acid. This information led to the proposal of structure **3.25** for mycelianamide. Labelling studies showed that it was assembled from an isoprene portion derived from mevalonate together with tyrosine and alanine as the amino acid units.

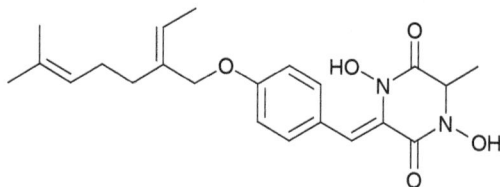

**3.25**

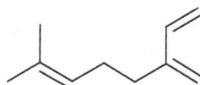

**3.26**

### 3.5.2 Gliotoxin

Gliotoxin is a highly anti-fungal and anti-bacterial metabolite of several *Gliocladium* and *Trichoderma* species such as *T. viride* as well as *Aspergillus fumigatus* and *Penicillium terlikowski*. It was first isolated in 1936 by Weindling and Emerson from *Gliocladium fimbricatum*. Gliotoxin is unusual in possessing a disulfide bridge. The key to its structure (**3.27**) involved the mild reduction with aluminium amalgam to a dethiogliotoxin, which could then be degraded to indole derivatives. The facile transformation of gliotoxin by alkaline alumina into the dehydro compound **3.28** led to the proposal in 1958 by Johnson and Woodward of the correct structure. The stereochemistry was established by X-ray crystallography in 1966.

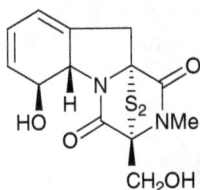

**3.27**

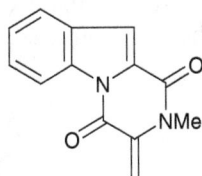

**3.28**

Some other fungal metabolites related to gliotoxin have been isolated, including some with S-methyl groups in place of the sulfur bridge. The S-methyl phenol **3.29** and its dimethylallyl ether have been isolated along with gliotoxin from *Gliocladium deliquescens*. In biosynthetic studies it has been shown that phenylalanine is incorporated into the indolecarboxylic acid portion of gliotoxin whilst serine provides the three-carbon unit. The N-methyl group was provided by the methyl group of methionine. The dipeptide, cyclo-(L-phenylalanyl-L-seryl) with the L-amino acid stereochemistry, was efficiently (48%) incorporated into gliotoxin whilst trapping experiments established the formation of the dipeptide by *Trichoderma viride*. The formation of gliotoxin may involve the cleavage of an arene oxide. Since the phenols that have been isolated as co-metabolites have a *para* rather than a *meta* relationship to the side-chain, a suggested pathway involves the arene oxide **3.30** rather than its isomer.

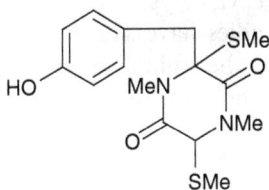

**3.29**

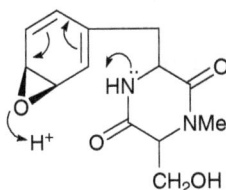

**3.30**

Despite its anti-bacterial activity against both Gram positive and Gram negative bacteria, gliotoxin was too toxic to be developed as an antibiotic.

However, gliotoxin also exhibits strong immunosuppressive activity. It has been shown to induce apoptosis of monocytes and it can selectively deplete bone marrow of mature lymphocytes. This is significant in that gliotoxin has been detected in the blood of patients suffering from invasive aspergilliosis. This infection by the fungus *Aspergillus fumigatus* is one of the most serious complications of immunocompromised patients.

The thiodiketopiperazine moiety has been found in other fungal metabolites such as aranotin (**3.31**) and the sporidesmins (**3.32**). The latter are the toxic metabolites of *Pithomyces chartarum*, a fungus that grows on pastures and causes facial eczema of cattle and sheep. In some metabolites such as dithiosilvatin and silvathione, obtained from *Aspergillus silvaticus*, both nitrogens of the diketopiperazine are methylated precluding the formation of an indole.

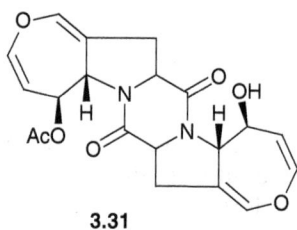

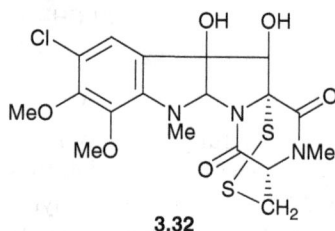

**3.31**          **3.32**

## 3.6 The Cyclopenin-Viridicatin Group of Metabolites

Anthranilic acid can form one component of the cyclic dipeptides that are metabolites of the genus *Penicillium*. The cyclopenin-viridicatin group are produced by *P. cyclopium*. They are inter-related by the following biosynthetic sequence (see Scheme 3.3). Methylation of the condensation product of anthranilic acid (**3.33**) and L-phenylalanine (**3.34**) affords cyclopeptine (**3.35**). Dehydrogenation gives dehydrocyclopeptine (**3.37**), which undergoes epoxidation to form cyclopenin (**3.36**). Rearrangement of the epoxide of cyclopenin or of a hydroxylation product, cyclopenol (**3.38**), affords the quinolines, viridicatin (**3.39**) and viridicatol (**3.40**). The enzyme systems that mediate these steps have been characterized. Labelling studies have shown that in the ring contraction of the benzodiazepine to form the quinoline ring the carboxyl group at C-5 is lost as carbon dioxide and the N-methyl group is split off as methylamine.

## 3.7 Tryptophan-derived Metabolites

The amino acid tryptophan provides the building block for some important metabolites, including the ergot alkaloids of *Claviceps purpurea* and the

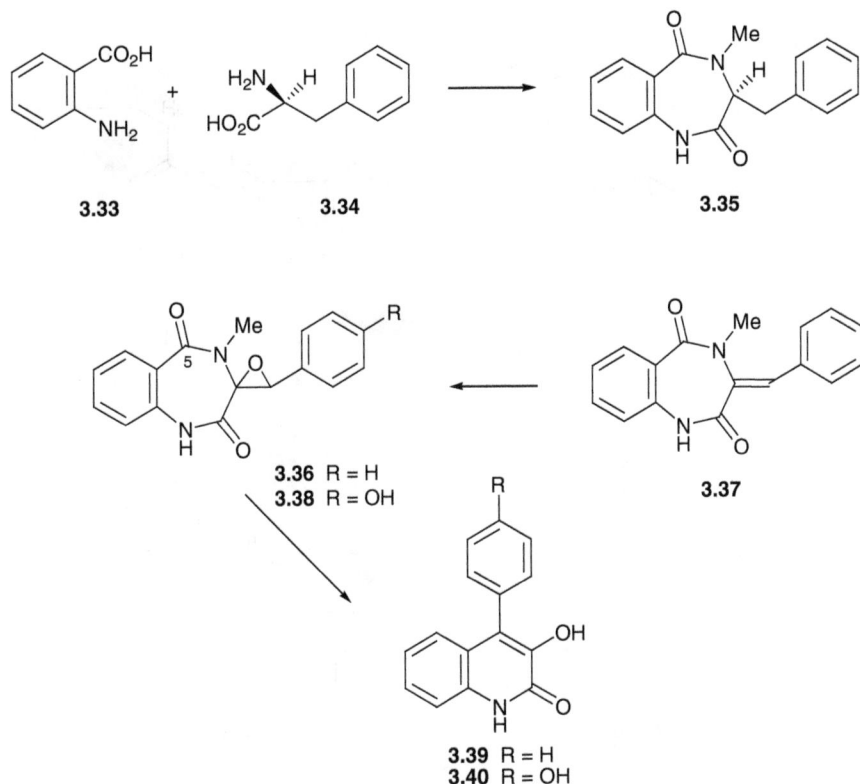

**Scheme 3.3** Relationship of the cyclopenin and viridicatin group of metabolites.

tremorgenic metabolites of *Penicillium roqueforti*. In man the indole amino acid tryptophan is a precursor of the neurotransmitter serotonin and thus there are receptors to which indoles will bind. Not surprisingly many of these fungal metabolites derived from tryptophan have neurotoxic properties. Consequently, they are discussed in Chapter 9 on mycotoxins. Tryptophan and phenylalanine together with a polyketide chain form the cytochalasins, which are discussed in Chapter 4 on polyketides.

Tryptophan and its relative indolylpyruvic acid (**3.42**) have been shown to precursors of hinnuliquinone (**3.41**), which is a pigment of *Nodulisporium hinnuleum*. Typical of many fungal indoles in which alkylation by a dimethylallyl or isopentenyl group has occurred, mevalonate was also a precursor. However, the stage at which prenylation of a monomer or a dimer took place was unclear. Asterriquinone (**3.43**) from *Aspergillus terreus* and cochliodinol (**3.44**) from *Chaetomium cochliodes* are similar metabolites. Fission of the hydroxyquinone in the latter followed by lactonization leads to cochliodinone (**3.45**) in a sequence that is similar to that which inter-relates the terphenyls and pulvinic acids described in Chapter 7.

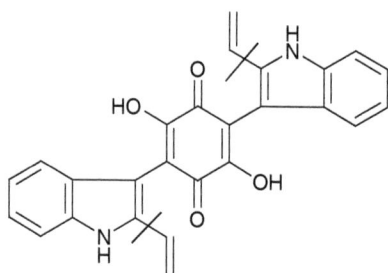

**3.41**

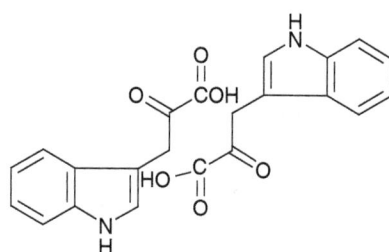

**3.42**

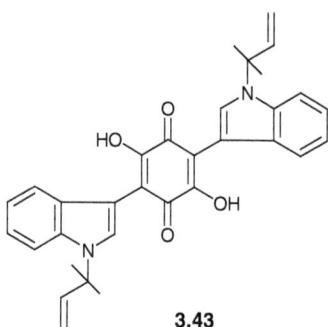

**3.43**

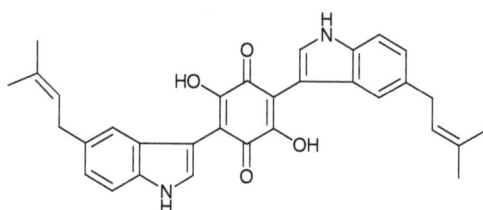

**3.44**

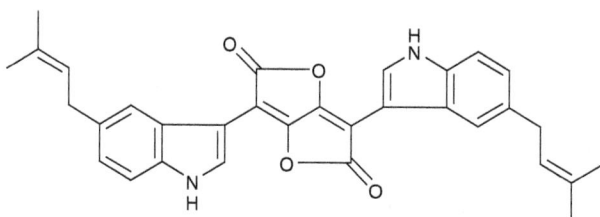

**3.45**

## 3.8 Glutamic Acid Derivatives

Glutamic acid and its derivatives have been found in several microorganisms. The common edible mushroom, *Agaricus bisporus*, contains γ-L-glutaminyl-4-hydroxybenzene (**3.46**) which is formed from 4-aminophenol and glutamic acid. This metabolite is readily hydroxylated to γ-L-glutaminyl-3,4-dihydroxybenzene and oxidized to the 3,4-benzoquinone. This *ortho*-quinone decomposes to an iminoquinone, 2-hydroxy-4-iminocyclohexa-2,5-dieneone (**3.47**), which imparts a red colour to some mushrooms.

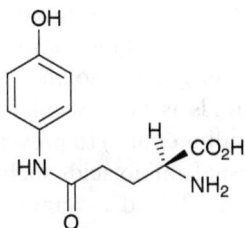

3.46

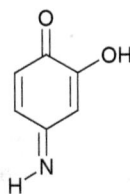

3.47

Agaratine (**3.48**) is another glutaminic acid derivative that is found in *Agaricus bisporus*, where it is present to the extent of 0.1–0.3% dry weight. This metabolite has attracted interest because it may be hydrolyzed enzymatically to 4-hydroxymethylphenylhydrazine and oxidized to the corresponding diazonium ion (**3.49**). The metabolic fate of agaritine has been linked with the carcinogenicity of some raw and baked mushrooms that has been observed in test animals. Agaritine appears to prevent melanin formation in mushrooms by reacting with *o*-quinones and it also inhibits the growth of *Trichoderma viride*, an organism with which it may be in competition in the wild.

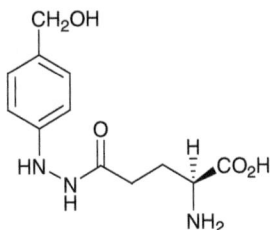

3.48

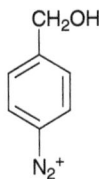

3.49

## 3.9 Fungal Peptides

Fungi produce several biologically active non-ribosomal peptides, some of which are toxic to man (see Chapter 9). A significant number of these are cyclic peptides, including macrocyclic lactones. Thus, the beauverolides and beauvericin are cyclic depsipeptides (peptido-lactones) that have been isolated from entomopathogenic species of *Beauveria bassiana*. Mass spectrometry played an important part in determining their structure. These compounds have a pathogenic effect on insects. Another series with insecticidal activity, and which are known as the destruxins, has been isolated from the insect pathogen *Metarrhizium anisopliae*.

HC-Toxin and victorin are produced by *Helminthosporium carborum*, which is a pathogen on maize, whilst another peptide, AM-toxin, is produced by *Alternaria alternata*, which causes a blotch disease on apples. The peptabiols, produced by *Trichoderma harzianum*, affect the development of other fungi.

Cyclosporins are a group of cyclic undecapeptides produced by *Tolypocla-dium inflatum* (*T. niveum* or *Beauveria nivea*). The major component is cyclo-sporin A. Although these metabolites were originally isolated as anti-fungal agents, the major interest in these compounds is because of their immuno-suppressant activity and their use (*e.g.* as Sandimmum®) to prevent rejection in transplant surgery. They include several unusual amino acids such as *N*-methyl-4(*R*)-4[(*E*)-2-butenyl]-4-methyl-L-threonine (**3.50**) and L-α-aminobutyric acid. Many of the amino acid residues, such as L-leucine and L-valine, are also present as their N-methyl derivatives whilst alanine is incorporated in the same cyclosporin as both the L- and D-enantiomer. An alanine racemase that converts L-alanine into D-alanine has been identified in the fungus. Analogues of cyclosporins have been synthesized by modifying the amino acid composition of the medium on which the fungus is grown.

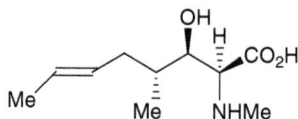

**3.50**

# CHAPTER 4
# *Polyketides from Fungi*

## 4.1 Introduction

Polyketides are natural products that are biosynthesized from poly-1,3-dicarbonyl compounds. The proposal that poly-1,3-diketones might form the biosynthetic precursors of aromatic compounds such as orsellinic acid was originally made by Collie in 1907 as a result of some laboratory experiments. Labelling studies by Birch in the 1950s on the biosynthesis of the fungal metabolites, 6-methylsalicylic acid and, subsequently, griseofulvin, provided experimental support for the theory. The formation of many aromatic compounds by this pathway may be seen as a combination of the biological equivalents of the Claisen and aldol condensations. A characteristic of these metabolites is oxygenation on alternate carbon atoms, *i.e.* on those atoms corresponding to the carbonyl groups of the precursor.

In fungi the poly-1,3-dicarbonyl precursors are formed from acetyl and malonyl co-enzyme A. Propionyl and butyryl co-enzyme A are also utilized by bacteria. The linear oligomerization of acetate units can lead to fatty acids and polyketides. Fungi utilize two related enzyme systems to carry out these biosyntheses, fatty acid synthase and polyketide synthase. However, although there is a formal similarity between these pathways, there is considerable evidence that even when they operate in the same organisms they do so with a different stereochemistry. Metabolites derived by the polyketide synthase pathway are commonly produced by the Ascomycetes and by the Fungi Imperfecti such as *Penicillium*, *Aspergillus*, *Fusarium* and *Alternaria* species but they are less common amongst the Basidiomycetes. The metabolites of polyketide synthase are common amongst the Streptomycetes.

Amongst the metabolites of *Aspergillus* and *Penicillium* species derived from acetate units are aromatic compounds, quinones, pyrones and lactones. They are grouped in terms of the number of acetate units incorporated into their carbon skeleton as tri-, tetra- and pentaketides, *etc*. Compounds containing four, five, seven and eight units are quite common but those with three, six, nine and ten or more acetate units that are formed by this pathway, are less

The Chemistry of Fungi
By James R. Hanson
© James R. Hanson, 2008

common. Extra units derived from methionine or the isoprene pathway may be introduced. Where such alkylations occur on the carbon chain, they normally do so at sites that are derived from the methylene of a polycarbonyl precursor. Several fungal metabolites are of mixed biosynthetic origin, part polyketide and part of amino acid or terpenoid origin.

Cyclization of the polycarbonyl chain to form aromatic compounds is a very common biosynthetic process. These aromatic compounds can then undergo various further biosynthetic transformations, including ring cleavage reactions. Some polyketides are pigments of fungi and others are serious mycotoxins and these are described in Chapters 7 and 9, respectively.

## 4.2   Polyketide Biosynthesis

There are parallels between fatty acid and polyketide biosynthesis. Whilst the starter unit for a polyketide chain is derived from acetyl co-enzyme A, the subsequent acetyl co-enzyme A units are activated for condensation by conversion into malonyl co-enzyme A. Over the last 20 years the chemical enzymology of this pathway has been studied in detail and some polyketide synthases have been isolated, mainly from Streptomycetes, particularly as a result of genetic studies by Hopwood. This chemical enzymology has been explored by several research groups, including those of Cane, Simpson and Staunton. The fungal 6-methylsalicylic acid synthase has also been examined in detail. Polyketide synthases are large highly organized multi-enzyme complexes with separate domains that function as an acyl transferase, a keto synthase, an acyl carrier protein, keto reductase, dehydratase, enoyl reductase and a thio-esterase. Each group of domains carry out transformations on an acetyl unit, although in different organisms not all are always expressed. Thus, the first stage (Scheme 4.1) in the biosynthesis of the fungal tetraketide 6-methylsalicylic acid (**4.1**) involves the transfer of an acetyl unit from an acetyl co-enzyme A to the keto synthase and a malonyl unit to the acyl carrier protein. The malonyl unit is formed by carboxylation of an acetate unit. Condensation of the acetate and malonate units takes place to give an acetoacetyl unit with the loss of the activating carboxyl as carbon dioxide. The acetoacetyl unit that is attached to the acyl carrier protein then migrates to the keto synthase and a further malonyl unit is attached to the acyl carrier protein. Another condensation occurs to give a triketone. In the case of 6-methylsalicylic acid biosynthesis, the central carbonyl group is reduced and the hydroxyl group is then eliminated prior to the addition of the fourth acetyl unit *via* malonyl co-enzyme A. When this unit has been added cyclization occurs and 6-methylsalicylic acid is released from the enzyme complex.

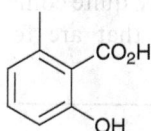

**4.1**

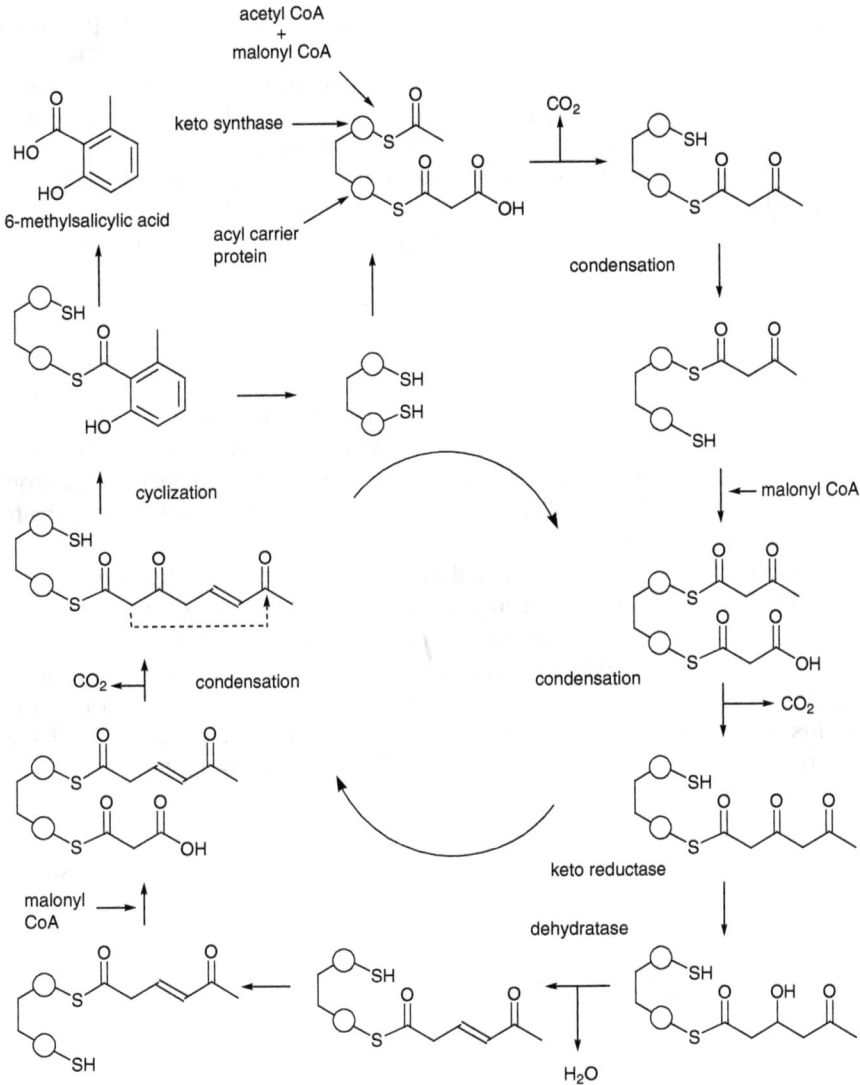

**Scheme 4.1**    Biosynthesis of 6-methylsalicylic acid.

As the polyketide chains become longer, there are sometimes several ways in which the polyketide chain may be folded to generate the same final structure. Thomas has pointed out that apparently similar decaketide aromatic compounds are formed by one folding of the polyketide chain in fungi and by a different folding in Streptomycetes.

Many *Penicillium* and *Aspergillus* species incorporate labelled acetate units into their polyketide metabolites sufficiently well to substantiate these biosynthetic proposals. Unfortunately, the nature of the polyketide synthase complex normally precludes the successful incorporation of short precursor polyketide chains into the developing chain. However, once the final chain and aromatic rings are formed, later intermediates may be successfully incorporated into the final metabolites. Because of this problem an underlying theme of this chapter is the ingenious way in which the results of various labelling experiments based on a simple two-carbon acetate precursor can be applied to elucidate biosynthetic pathways. The use of carbon-13 labelling and NMR methods of detection have been particularly important in this connection.

## 4.3    Triketides

Triketides are relatively rare. Triacetic acid lactone (**4.2**) has been detected in *Penicillium patulum*. It is also produced by fatty acid synthase in the absence of the reductant NADPH. Radicinin (**4.3**) is a major phytotoxin isolated from *Alternaria radicina (Stemphyllium radicinum)* which causes a black rot of carrots. It is also formed by other *Alternaria* species. Its pyrano[4,3-*b*]pyran structure, the identification of which had eluded purely chemical degradative studies, was established in one of the earlier applications of $^1$H NMR spectroscopy to natural product structure elucidation. The biosynthesis of radicinin from acetate units was studied in 1970 by both radio-isotope methods using carbon-14 and by carbon-13 enrichment studies with NMR methods of detection. This was one of the first applications of this NMR technique to biosynthetic problems. These results established the labelling pattern for radicinin shown in **4.3**.

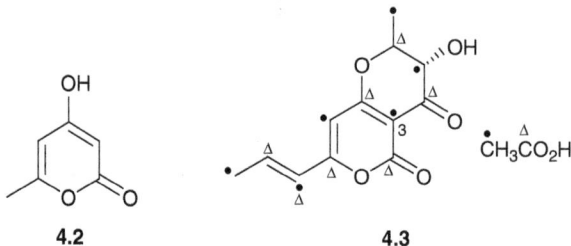

**4.2**                               **4.3**

Radicinin may be constructed from two polyketide chains. Since good incorporations of acetate were observed in these fermentations, a revealing experiment could be carried out using the fungus *Alternaria helianthi*. The fungus was grown on a medium that contained an equimolar amount of [1-$^{13}$C]- and [2-$^{13}$C]acetate so that, when the polyketide chain was assembled, there was a chance that two labelled units, one from [l-$^{13}$C]acetate and the other from [2-$^{13}$C]acetate, would adjoin each other. The resultant metabolite would show a coupling from these adjacent labelled units. The C-3 of radicinin and deoxyradicinin appeared as a doublet of doublets in the $^{13}$C NMR spectrum, showing that it was at the junction of two polyketide chains. The results of this experiment contrast with that of feeding [1,2-$^{13}$C$_2$]acetate, which gives information on the incorporation of intact acetate units. When radicinin was biosynthesized from [2-$^2$H$_3$]acetate each of the terminal methyl groups retained three deuterium atoms and, therefore, they represent the starting points of two polyketide chains. Radicinin may be constructed from two C$_6$ chains or a C$_9$ and a C$_4$ chain.

Labelling studies using $^{13}$C NMR methods have shown that the dilactone colletodiol (**4.4**), which is produced by *Colletotrichum capsici*, is formed from a tri- and a tetraketide chain. These dilactones are discussed later.

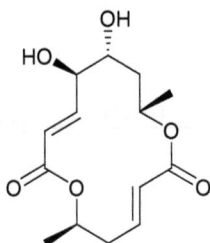

**4.4**

# 4.4   Tetraketides

### 4.4.1   6-Methylsalicylic Acid

6-Methylsalicylic acid (**4.1**) is the archetypical tetraketide. The study of its biosynthesis and relationship to other fungal metabolites has been a central part of polyketide biosynthesis. Biosynthetic alkylation, decarboxylation, further hydroxylation and ring cleavage reactions of 6-methylsalicylic acid and its related dihydroxyphenol, orsellinic acid, afford many fungal metabolites. 6-Methylsalicylic acid has been isolated from several *Penicillium* species. It was first detected in 1930 by Raistrick in the culture filtrate of *Penicillium griseofulvum* as a result of the intense violet colour that it gave with a ferric chloride solution and the precipitate that was formed when the culture filtrate was treated with bromine water. It was identified by comparison of its melting point and that of its methyl ester with the literature data of synthetic material. An authentic sample was also synthesized from 2-amino-3-nitrotoluene (**4.5**). 6-Methylsalicylic acid showed anti-bacterial activity.

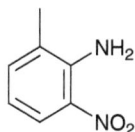

**4.5**

In 1955, 6-methylsalicylic acid was the first fungal metabolite to be used to test the acetate hypothesis for the biosynthesis of aromatic compounds. The original proposal had been made by Collie in 1907 and then developed by Birch in 1953. This experiment, like many early biosynthetic studies, was carried out by

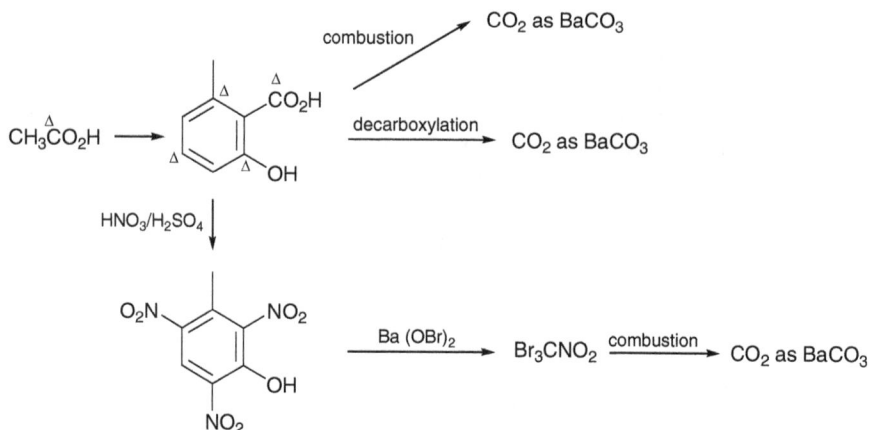

**Scheme 4.2**   Degradation of labelled 6-methylsalicylic acid.

growing the fungus in a medium containing radioactive sodium [l-$^{14}$C]acetate. The resulting 6-methylsalicylic acid had to be degraded to establish the specificity of the label. The degradation is set out in Scheme 4.2. This sequence illustrates the disadvantages of the older radio-labelling experiments and the advantages that accrue from the more modern NMR methods based on carbon-13 and deuterium labelling, which avoid the need for extensive chemical degradation. The enzymatic steps that lead to the formation of 6-methylsalicylic acid have been described above (Section 4.2).

Biosynthetic modifications of 6-methylsalicylic acid (**4.1**) and orsellinic acid (**4.6**) occur in several fungi to generate further tetraketide metabolites. Decarboxylation of these compounds followed by hydroxylation and oxidation leads to a family of toluquinones. These are widespread as metabolites of *Penicillium* and *Aspergillus* species. They are exemplified by fumigatin (**4.7**) and spinulosin (**4.8**), which were isolated from *Aspergillus fumigatus*.

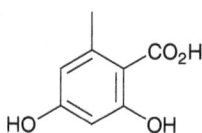

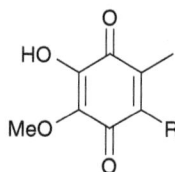

**4.6**                                         **4.7**  R = H
                                                **4.8**  R = OH

## 4.4.2   Patulin and Penicillic Acid

Two interesting metabolites of *Penicillium patulum* are patulin and penicillic acid. Their biosynthesis involves the cleavage of an aromatic ring. These substances are mycotoxins and their activity in this context is discussed in Chapter 9.

Patulin (**4.9**) was isolated from *Penicillium patulum* in 1943. Since it had a quite marked anti-bacterial activity it was detected and then isolated from several other *Penicillium* and *Aspergillus* species during the early 1940s. It is a particular hazard as a metabolite of *P. expansum*, a spoilage organism that grows on apples. It proved to be too toxic to be of any use as an antibiotic. Although patulin has a simple molecular formula, $C_7H_6O_4$, its structure proved quite difficult to establish by the purely chemical means available at the time. The carbon skeleton of patulin was established by hydrogenation and by further reduction with hydrogen iodide and red phosphorus to form 3-methylhexanoic acid and 3-methyl-4-hydroxyhexanoic acid 1,4-lactone (**4.10**). 6-Iodo-4-ketohexanoic acid was formed in another sequence, involving reduction and hydrolysis with hydrogen iodide. Acid hydrolysis also gave formic acid and a low yield of tetrahydro-4-pyrone-2-carboxylic acid (**4.11**). This decomposition suggested the presence of a masked β-formylketone in the structure. Earlier formulations for patulin were eventually corrected by Woodward in 1949, who synthesized deoxypatulin and, in 1950, patulin itself, using tetrahydro-4-pyrone as a starting material.

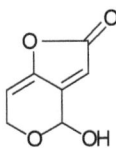 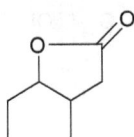 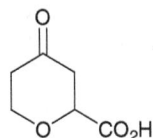

    **4.9**          **4.10**          **4.11**

It is interesting to speculate how quickly the structure might have been determined had NMR spectroscopy been in widespread use at the time. Indeed infrared correlations, which were being developed at the time of Woodward's proposal for the structure of patulin, were in accord with the structure. The presence of the enol of a ketone and a ketal in the structure also play an important part in the biological activity of patulin.

In 1953, Birkinshaw suggested that patulin might be formed from an aromatic compound and in 1958 it was shown by Bu'Lock that 6-methylsalicylic acid (**4.1**) fulfilled this role (see Scheme 4.3). The incorporation of 6-methylsalicylic acid into patulin involved the loss of the carboxyl group. The experiment was carried out with 6-methylsalicylic acid that had been prepared biosynthetically from [1-[14]C]acetate and the patulin was degraded to establish the sites of the remaining three acetate labels. Using deuteriated precursors and mass spectroscopic analysis, Scott was able to show that *m*-cresol (**4.12**), formed by decarboxylation of 6-methylsalicylic acid, was hydroxylated to **4.13** and converted *via* gentisyl alcohol (**4.14**) into patulin (**4.9**). The post-gentisyl alcohol sequence was established using mutants of *P. patulum* and *P. urticae* that were blocked at various stages in patulin biosynthesis. This led to the identification of the quinone (**4.17**) and the epoxide intermediates, isoepoxydon (**4.15**) and phyllostine (**4.16**). An isomer of patulin (isopatulin, **4.19**) was formed first and then converted into patulin *via* the diol ascladiol (**4.18**).

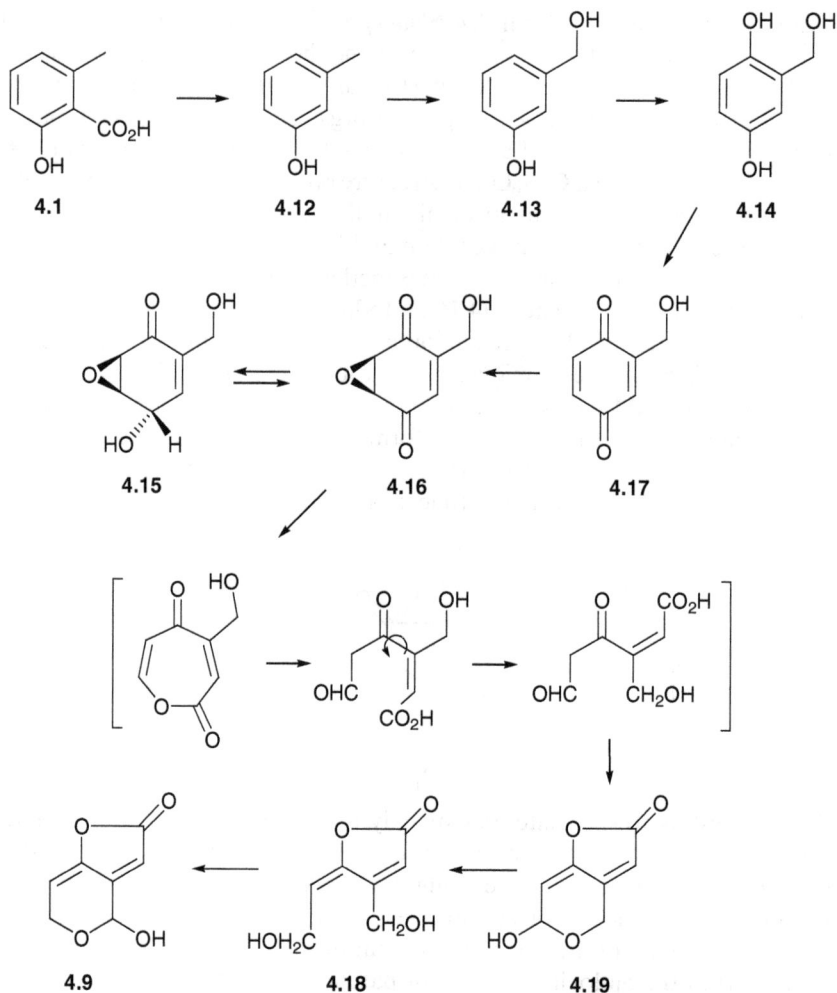

**Scheme 4.3**   Biosynthesis of patulin.

The biosynthesis of patulin was also studied using [l-$^{13}$C, $^{18}$O$_2$]acetate and $^{18}$O$_2$ gas and then measuring the oxygen-18 induced chemical shifts in the $^{13}$C NMR spectrum to locate the site of the $^{18}$O. This showed that only the carbonyl oxygen was derived from acetate and that the others were derived by oxidative processes. Many of the enzyme systems that mediate these steps have been isolated from *P. patulum*.

Penicillic acid (**4.23**) was first isolated in 1913 from *Penicillium puberulum* and subsequently from *P. cyclopium* in 1936 by Raistrick who established its structure. Its derivation from orsellinic acid (**4.20**) was suggested by Birch in 1958 as a result of feeding experiments with [1-$^{14}$C]acetate. The labelling pattern from acetate was also determined in 1960 by Mosbach, who suggested an alternative cleavage of the aromatic ring (see Scheme 4.4). The use of coupling patterns derived from

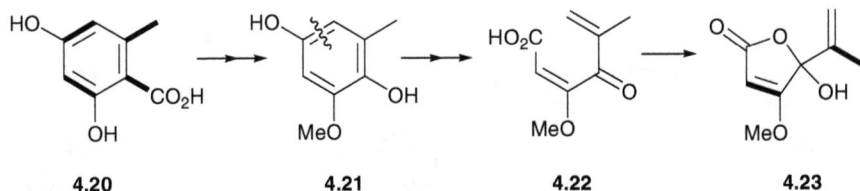

**Scheme 4.4** Biosynthesis of penicillic acid.

feeding doubly-labelled [1,2-$^{13}C_2$]acetate to distinguish between plausible biosynthetic pathways is well illustrated here. Orsellinic acid, biosynthesized from [1,2-$^{13}C_2$]acetate, appropriately diluted with unlabelled material to avoid interunit coupling, would be labelled as shown in **4.20**. The decarboxylation and ring cleavage proposed by Birch, involving fission of the bond between the hydroxyl and carboxyl group, would lead to penicillic acid retaining three intact acetate units and hence three pairs of $^{13}C-^{13}C$ couplings. The alternative Mosbach pathway *via* **4.21** and **4.22** would retain only two intact units and hence only two pairs of $^{13}C-^{13}C$ couplings. The presence of only two groups of couplings in the penicillic acid biosynthesized from [1,2-$^{13}C_2$]acetate confirmed that the Mosbach pathway (Scheme 4.4) was involved in the biosynthesis. There was a randomization of the coupling between the methyl and the methylene group. Scission of the 4,5-bond of an orsellinic acid precursor was also supported in work by Elvidge that made a rare use of $^3H$ NMR spectroscopy in biosynthetic studies.

### 4.4.3 Gladiolic Acid and its Relatives

The alkylation of these aromatic rings by methyl and prenyl groups generates further fungal metabolites. Gladiolic acid (**4.24**) and the corresponding primary alcohol, dihydrogladiolic acid (**4.25**), are produced by *Penicillium gladioli*. The labelling patterns from [1-$^{13}C$]-, [1,2-$^{13}C_2$]- and [2-$^2H_3$]acetates and from [Me-$^2H_3$]methionine established the origin of the atoms as shown in **4.25**. In this metabolite, as with many others, extra carbon atoms such as methyl groups have their origin in the methyl group of methionine.

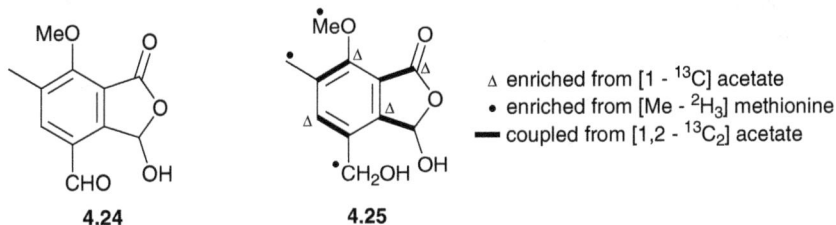

Δ enriched from [1 - $^{13}C$] acetate
• enriched from [Me - $^2H_3$] methionine
— coupled from [1,2 - $^{13}C_2$] acetate

Other examples are flavipin (**4.26**) from *Aspergillus flavipes*, quadrilineatin (**4.27**) from *A. quadrilineatus* and cyclopaldic and cyclopolic acids (**4.28**) from *Penicillium cyclopium*.

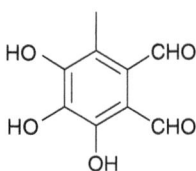

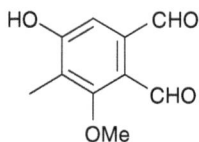

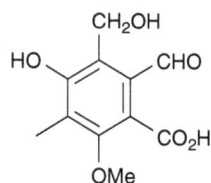

**4.26**                              **4.27**                              **4.28**

The presence of a vicinal dialdehyde in flavipin (**4.26**) was revealed by the formation of a phthalazine with hydrazine. The substitution pattern of the aromatic ring was established by treatment of the trimethyl ether with potassium hydroxide. The dialdehyde underwent an internal Cannizzaro reaction to form a phthalide. This phthalide was synthesized by bis-chloromethylation of the trimethyl ether of gallic acid and hydrogenolysis of the chlorine. The biosynthesis of flavipin from acetate units and methionine and from aromatic precursors showed that it was a tetraketide in which methylation had taken place after the polyketide chain had been formed. Flavipin is active against a wide range of plant pathogenic fungi. It has been isolated, along with some carotenoids, as a metabolite of *Epicoccum nigrum*. This is a red spoilage organism found on agricultural products. Interest has been shown in this compound because of its activity against *Monolinia laxa*, which is the causative organism of peach twig blight. These aromatic orthodialdehydes inhibit electron transport and oxidative phosphorylation in plant mitochondria.

### 4.4.4  Tetraketide Tropolones

The fungal tropolones puberulic acid (**4.29**) and stipitatic acid (**4.30**) were isolated from *P. puberulum* and *P. stipitatum* by Raistrick in 1932 and 1942, respectively. Their structures were not established at that time despite extensive chemical degradation. The tropolone structure for stipitatic acid was proposed by Dewar in 1945 and that for puberulic acid in 1950 by Todd. These structural proposals played an important part in the development of ideas of aromaticity. The biosynthesis of these tropolones involves the ring expansion of an orsellinic acid *via* stipitatonic acid (**4.31**).

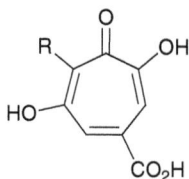

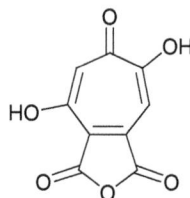

**4.29**  R = OH
**4.30**  R = H                                              **4.31**

## 4.4.5 Mycophenolic Acid

Although it may be considered as a meroterpenoid (see Chapter 5) mycophenolic acid (**4.32**) is discussed here because its structure reveals the interaction between polyketides and terpenoids. A metabolite of a *Penicillium* species, which was probably mycophenolic acid, was first described by Gosic in 1896. It was more firmly characterized by Alsberg and Black in 1913 as a metabolite of *P. stolonifer*. Mycophenolic acid was obtained in 1931 from *P. brevi-compactum* by Raistrick who proposed a tentative structure in 1948, which was confirmed in 1952. The structure of the phthalide was established by fusion of the metabolite with potassium hydroxide to form 1,5-dihydroxy-3,4-dimethylbenzene (**4.33**). Potassium permanganate oxidation of the methyl ether of mycophenolic acid gave the keto-acid **4.34**, which on further degradation gave the known anhydride **4.35**. Cleavage of the alkene in the side chain by ozonolysis gave levulic acid and an aldehyde that was shown to be **4.36**. These results (Scheme 4.5) together with other degradative evidence led to the structure **4.32** for mycophenolic acid.

Biosynthetic studies showed that the aromatic ring was formed from a polyketide whilst the side chain was derived from the mevalonate terpenoid pathway. The C-5 O-methyl and C-4 methyl groups were derived from methionine. The high incorporation of 4,6-dihydroxy-2,3-dimethyl-[l-$^{14}$C]benzoic acid (**4.37**) suggested that the methylation at C-4 occurred at the polyketide stage prior to the formation of the phthalide. 5,7-Dihydroxy-4-methyl-[7-$^{14}$C]phthalide (**4.38**) and its 6-farnesyl analogue were also incorporated efficiently into mycophenolic acid. The farnesyl precursor was detected as a metabolite of *P. brevi-compactum* and shown to be formed from 5,7-dihydroxy-4-methyl-[7-$^{14}$C]phthalide by a trapping

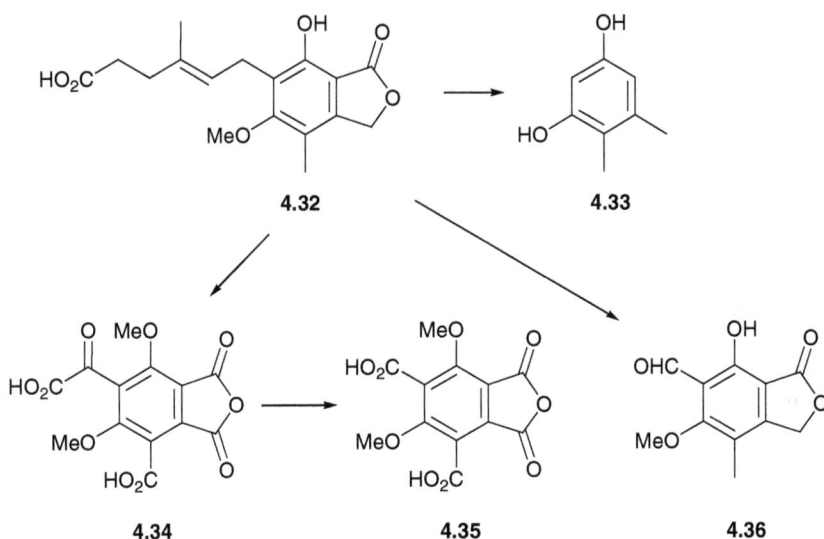

**Scheme 4.5** Degradation of mycophenolic acid.

experiment. It was a much more efficient precursor than the corresponding geranyl derivative.

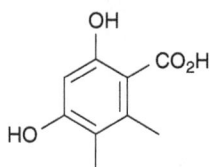                    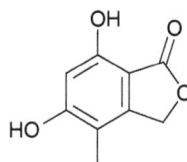

**4.37**                                    **4.38**

Mycophenolic acid has anti-fungal, anti-bacterial and anti-viral activity. It is a potent inhibitor of inositol monophosphate dehydrogenase and guanosine monophosphate synthase and hence of nucleotide biosynthesis. Mycophenolic acid has a powerful immunosuppressive activity and the morpholine amide (CellCept®) is used to prevent rejection of organ transplants and has been recommended for the treatment of psoriasis.

## 4.5   Pentaketides

Many of the structural variations found in aromatic fungal metabolites that are tetraketides are also found in the pentaketides that are biosynthesized by *Penicillium* and *Aspergillus* species. Several fungal metabolites have a carbon skeleton that is derived from acetylorsellinic acid. This compound is itself a metabolite of *Penicillium brevi-compactum*. Cyclization of acetylorsellinic acid (**4.39**) forms isocoumarins such as mellein (**4.40**). Additional carbon atoms are inserted from methionine.

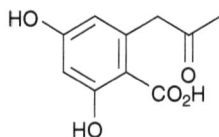                    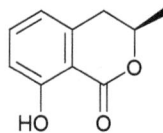

**4.39**                                    **4.40**

### 4.5.1   Citrinin

Citrinin (**4.41**) is a mycotoxin that is formed in quite large amounts by *Penicillium citrinum*, where it co-occurs with related metabolites such as the isocoumarins **4.42** and **4.43**, the phenols **4.44** and **4.45**, and decarboxycitrinin. Some dimers have also been isolated. Citrinin was first isolated by Raistrick in 1931 and its structure was proposed by Robertson in 1948. Although it exists as a quinone-methide in the crystalline state, it forms a hydrate rather rapidly. Much of the chemistry of citrinin can be understood in terms of the ready formation of this hydrate. The lactol of citrinin hydrate is readily oxidized to a lactone. The hydrate is a typical phenolic acid and undergoes decarboxylation whilst the lactol ring can be opened to a hydroxy-aldehyde. This can easily lose formaldehyde. The product, phenol A (**4.44**), was identified by synthesis.

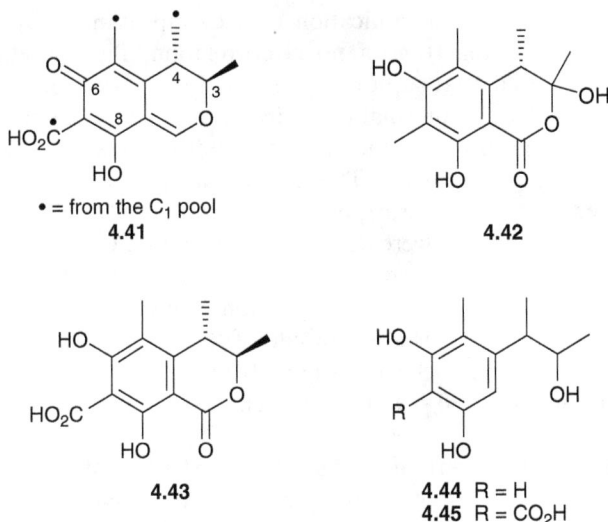

• = from the $C_1$ pool
**4.41**

**4.42**

**4.43**

**4.44** R = H
**4.45** R = $CO_2H$

Citrinin (**4.41**) is biosynthesized from five acetate units and three $C_1$ units derived from methionine or formic acid. The labelling experiments were carried out with $[1-^{14}C]$acetate and $[^{14}C]$formic acid. Degradation of the resultant citrinin established that the two methyl groups and the carboxyl group were introduced onto a straight acetic acid derived chain. In the original experiment the $C_1$ pool was labelled using formic acid. Evidence has been presented to show that the methylation of the polyketide chain takes place whilst it is still enzyme-bound. The first enzyme-free aromatic intermediate is the substituted benzaldehyde **4.46**. In an interesting experiment citrinin was biosynthesized from glucose in a deuterium oxide medium. This experiment showed that the hydrogen atom at C-4 was derived from the C–H of a glucose and hence established the oxidation of level of the isocoumarin intermediate. Other experiments were based on feeding multiply-labelled acetate precursor and confirmed these observations. Isotopically shifted $^{13}C$ NMR signals for C-3, C-6 and C-8 in material biosynthesized from $[1-^{13}C,^{18}O_2]$acetate indicated that the oxygen atoms at these centres were derived from acetate.

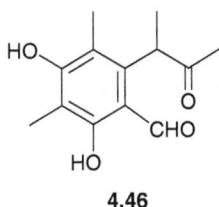

**4.46**

## 4.5.2 Terrein

Terrein (**4.47**) is another pentaketide fungal metabolite with an interesting biosynthesis. Although the metabolite was first isolated by Raistrick in 1935 from *Aspergillus terreus*, its structure was not established until 1954 by Grove. This study hinged on the identification of a cyclopentanone by IR spectroscopy, which was at that time a novel correlation. Birch studied the biosynthesis and showed that although the compound was formed from five acetate units there was an unusual break in the alternating pattern of the label from the carboxyl group of acetate such that both C-6 and C-7 were derived from the carboxyl group (**4.48**). The pattern of incorporation of [1, 2-$^{13}$C$_2$] acetate showed that terrein incorporated three intact C$_2$ units whilst the remaining two carbon atoms were derived from separate C$_2$ units. Studies with advanced precursors established that the dihydroisocoumarin **4.49** was specifically incorporated without prior degradation to acetate units. The dihydroisocoumarin is also known as a metabolite of *Aspergillus terreus*.

A biosynthetic scheme (Scheme 4.6) involving an oxidative decarboxylation and ring contraction (**4.50**→**4.51**) has been devised to accommodate these results.

The rearrangement of carbon chains with the extrusion of a carbon atom has been identified in several other biosyntheses. An example is provided by aspyrone (**4.52**) and asperlactone (**4.53**), which are metabolites of *Aspergillus melleus*. The labelling pattern from [1,2-$^{13}$C$_2$]acetate and from individually labelled acetates suggested that the carbon skeleton was formed from a linear polyketide chain that had undergone a rearrangement and decarboxylation (**4.54**). Labelling with oxygen-18 indicates that none of the oxygen atoms were derived from acetate units but that they arose from oxygen of the air. A biosynthetic scheme (Scheme 4.7) involving the intervention of epoxide intermediate **4.55** has been proposed to accommodate these results. A different way of opening the epoxide accounts for the formation of asperlactone (**4.53**).

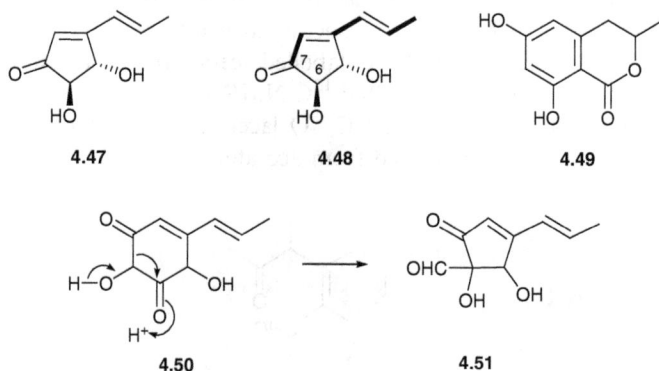

**Scheme 4.6**  Biosynthesis of terrein.

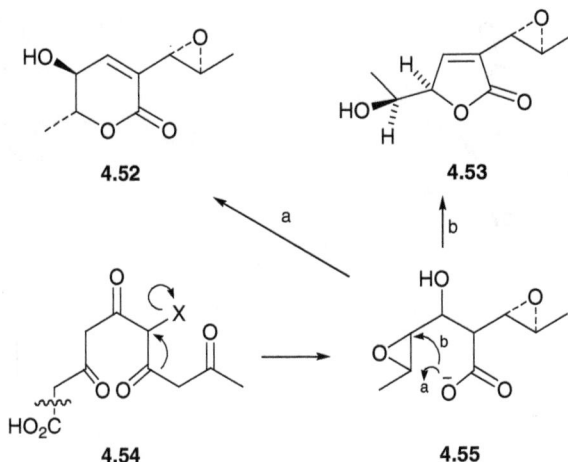

**Scheme 4.7** Biosynthesis of aspyrone and asperlactone.

Harzianolide (**4.56**) is an anti-fungal metabolite of *Trichoderma harzianum*. It contains a butenolide ring and labelling studies have suggested that it may be formed by a similar process.

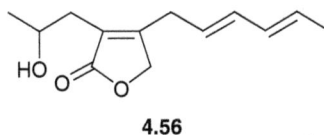

**4.56**

## 4.6   Hepta- and Octaketides

Aromatic compounds derived from six, seven and eight or more acetate units are also found as metabolites of these fungi. These metabolites include a series of anthraquinone pigments, which are discussed later in Chapter 7.

### 4.6.1   Griseofulvin

Griseofulvin (**4.57**) is an important heptaketide that is used as an anti-fungal agent (Fulcin® or Grisovin®). It was originally isolated in 1939 by Raistrick along with mycelianamide (see Chapter 3) from the mycelium of *Penicillium griseofulvum*. In 1946 it was obtained by Brian from *P. janczewski* as an anti-fungal agent that had an unusual curling and stunting effect on the hyphae of susceptible organisms such as the phytopathogen *Botrytis allii*. Most of the susceptible organisms contain chitin in their cell walls. Griseofulvin has a useful anti-fungal activity against organisms that are responsible for diseases of the skin in man such as ringworm.

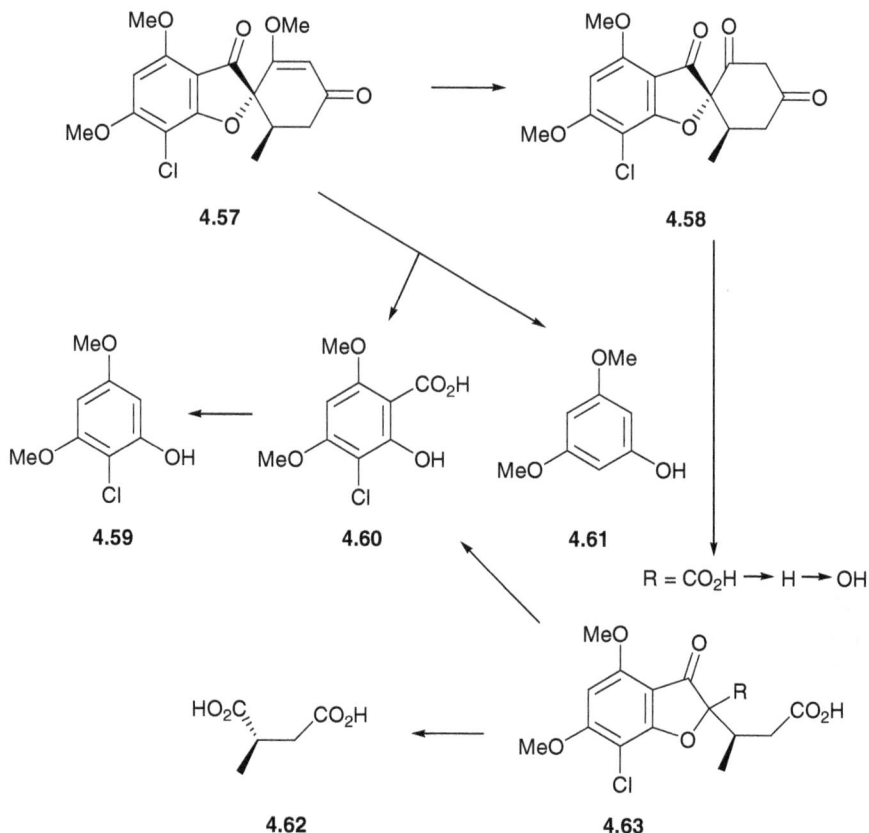

**Scheme 4.8** Degradation of griseofulvin.

The structure of griseofulvin was established in 1952 by a group of workers at the Akers Research Laboratories of ICI. Other aspects of the chemistry were explored by a group at Glaxo. Three features contribute to understanding the degradative chemistry (see Scheme 4.8) involved in the structure elucidation of griseofulvin. The first is the presence of a vinylogous β-diketone in the heart of the molecule, which enabled a retro-Claisen cleavage to take place. This afforded two fragments, **4.60** and **4.61**, which were identified and accounted for all the carbon atoms in the molecule. The second feature is the presence of a β-methoxy-α,β-unsaturated ketone, which was readily hydrolysed to a weakly acidic β-diketone, griseofulvic acid (**4.58**). The vinyl ether of griseofulvin was sensitive to oxidation. The third feature is the presence of an aryl chloride. The presence of chlorine in the degradation products, *e.g.* **4.59**, identified those products originating from the aromatic ring of griseofulvin. Thus, vigorous alkaline degradation gave the acid **4.60** and orcinol (3,5-dihydroxytoluene) monomethyl ether (**4.61**). The acid **4.60**, typical of a 2-hydroxybenzoic acid, lost carbon dioxide on heating to give 2-chloro-3,5-dimethoxyphenol (**4.59**),

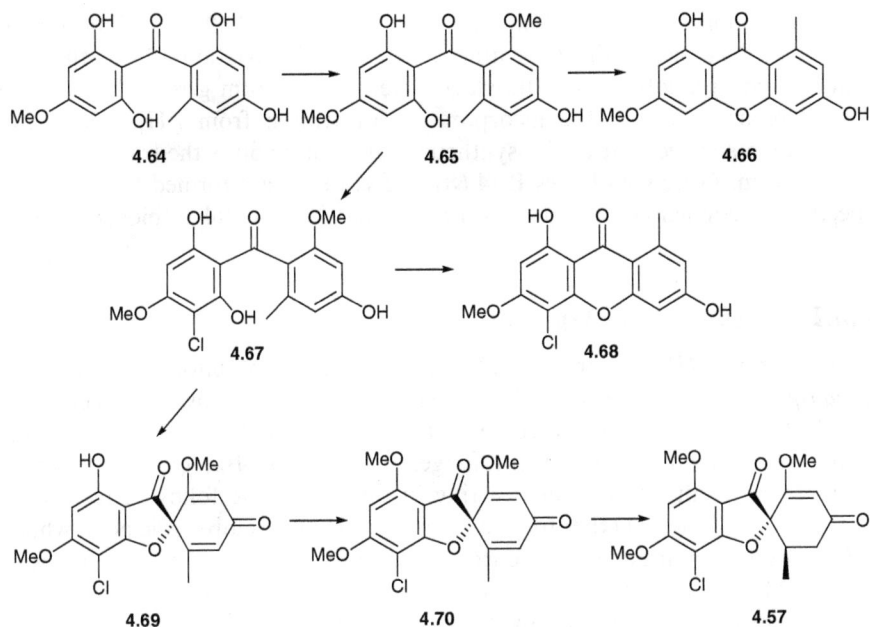

**Scheme 4.9** Biosynthesis of griseofulvin.

which was identified by synthesis. A systematic stepwise oxidative degradation of griseofulvic acid *via* **4.63** eventually gave the same acid and (+)-methyl-succinic acid (**4.62**). This degradation not only served to establish the structure of the non-benzenoid carbocyclic ring but also, in the formation of (+)-methylsuccinic acid, provided evidence for the absolute stereochemistry. The overall structure was confirmed by an X-ray crystallographic analysis.

The biosynthesis of griseofulvin (Scheme 4.9), which illustrates several general biosynthetic reactions, is one of the classical studies of biosynthesis. The hepta-ketide origin of griseofulvin was demonstrated by Birch in 1958 using [1-$^{14}$C] acetate and then confirmed by Tanabe in 1966 using [2-$^{13}$C]acetate. A further study by Simpson of the incorporation of singly- and doubly-labelled [$^{13}$C]acetate confirmed the folding of the heptaketide chain. The $^{13}$C NMR signals arising from the aromatic ring showed two distinct sets of $^{13}$C–$^{13}$C couplings arising from two different distributions of the acetate units around the ring. This indicated that, once the polyketide chain was formed, there was a symmetrical intermediate that could undergo free rotation. The incorporation of oxygen-18 from [1-$^{13}$C,$^{18}$O$_2$] acetate, determined by isotope shift $^{13}$C NMR experiments, indicated that all the oxygen atoms of griseofulvin were derived from the acetate units.

Griseofulvin is accompanied by the griseophenones A–C and by the gri-seoxanthones B and C. The role of the griseophenones in the biosynthesis was established by labelling and dilution analysis. In this pathway the benzophenone **4.64** is methylated to griseophenone C (**4.65**) and then chlorinated to form griseophenone B (**4.67**). The formation of the grisan carbon skeleton arises by a phenol coupling reaction of **4.67** to form the dienone **4.69**, providing one of the classic examples of this biosynthetic reaction. Methylation affords the dienone dehydrogriseofulvin (**4.70**), which is reduced to form griseofulvin (**4.57**). The stereochemistry of the incorporation of tritium from [$^3$H]acetate into griseofulvin showed that the biosynthetic reduction involves the trans addition of hydrogen. Griseoxanthones B (**4.68**) and C (**4.66**) are formed from the respective griseophenones but play no further part in griseofulvin biosynthesis.

## 4.6.2  Cladosporin (Asperentin)

Cladosporin (**4.71**) is an anti-fungal and plant growth inhibitory metabolite of *Cladosporium cladosporioides*. The same compound, also known as asperentin, was isolated independently together with its 6- and 8-O-methyl ether and its 5′-hydroxy derivative from an entomogenous strain of *Aspergillus flavus*. The presence of a dihydroisocoumarin ring in the metabolite followed from oxidation of the dimethyl ether to give 2,4-dimethoxy-6-oxalobenzoic acid, whilst NMR studies confirmed the presence of the tetrahydropyran. The structure was confirmed by X-ray analysis of a sample isolated from *A. repens*. The absolute stereochemistry was assigned by comparison of the circular dichroism curve with that of (*R*)-(−)-mellein.

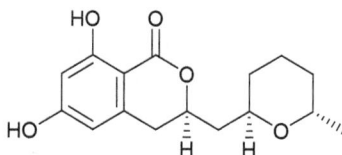

**4.71**

The stereochemistry of the biosynthesis has attracted interest in a comparison with fatty acid biosynthesis in the same organism. Studies with sodium [1-$^{13}$C]-, [2-$^{13}$C]-, [1,2-$^{13}$C$_2$]-, [1-$^{13}$C, $^{18}$O$_2$]-, [1-$^{13}$C, $^2$H$_3$]- and [2-$^{13}$C, $^2$H$_3$]acetate showed that cladosporin was an octaketide in which all of the oxygen atoms had their origin in the acetate units. Where the oxygen of an acetate had been retained in cladosporin, and there was a methylene arising from the following acetate unit. Deuterium labelling established that both hydrogens of the methylene were derived from acetate. However, where elimination of the oxygen had taken place followed by a biosynthetic reduction of an enoyl thiol ester, only one deuterium remained. It was shown that this remaining hydrogen had taken up the (*S*)-configuration in the developing polyketide chain. This configuration was opposite to that observed for fatty acid biosynthesis in the same organism. Furthermore, the reduction of the carbonyl groups in the polyketide chain gave alcohols with an (*S*)-configuration, which is opposite to the stereochemistry of reduction in fatty acid biosynthesis. These stereochemical differences between polyketide and fatty acid biosynthesis have been observed with other fungal metabolites such as dehydrocurvularin (**4.72**).

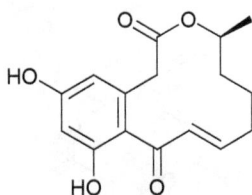

**4.72**

## 4.7 Polyketide Lactones

The biosynthetic reduction of the polyketide chain may prevent the formation of aromatic products. Instead, several lactones are formed. For example, a series of bio-active pentaketide metabolites containing a ten-membered ring have been isolated. These include the diplodialides (*e.g.* diplodialide A, **4.73**) from *Diplodia pinea*, the pyrenolides (*e.g.* pyrenolide A, **4.74**) from *Pyrenophora teres* and the cephalosporolides (*e.g.* cephalosporolide B, **4.75**) from *Cephalosporium aphidicola*. An unusual dimeric pentaketide, thiobiscephalosporolide A (in which the two lactones are held together by a sulfur), has also been obtained from *C. aphidicola*. Other metabolites with a ten-membered lactone ring include the decarestrictins (**4.76**) from *Penicillium* species such as *P. simplicissimum* and putaminoxin (**4.77**), which is a phytotoxic metabolite of *Phoma putaminum*. Recifeiolide (**4.78**) from *C. recifei* is a hexaketide with a 12-membered ring. Cladospolide A (**4.79**), also with a 12-membered ring, is a phytotoxic metabolite of *Cladosporium fulvum*. The decarestrictins affect cholesterol biosynthesis whilst the diplodialides were found to be steroid 11-hydroxylase inhibitors. Biosynthetic experiments with [$^{13}$C]-labelled acetate established that the decarestrictins were polyketides.

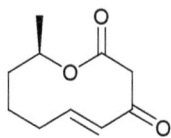

**4.73**

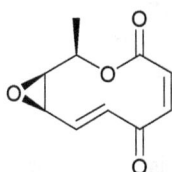

**4.74**

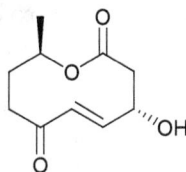

**4.75**

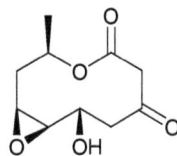

**4.76**

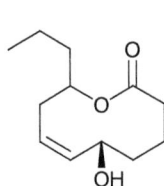

**4.77**

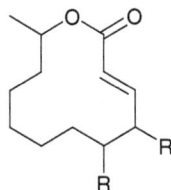

**4.78** R = H
**4.79** R = OH

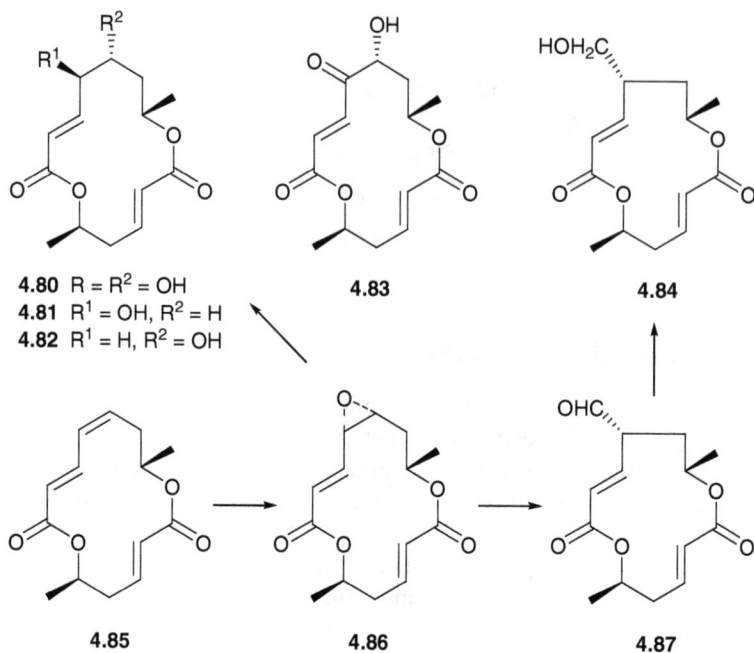

**Scheme 4.10**  Biosynthesis of the macrodiolides of *Cytospora* sp.

A series of polyketide macrodiolides (bislactones) have been isolated from fungi. Colletodiol (**4.80**), which contains a 14-membered ring, was first isolated from the plant pathogen *Colletotrichum capsici*, and shown to consist of a tri-ketide and a tetraketide chain. Several related macrodiolides have also been isolated, including colletallol (**4.81**), colletol (**4.82**) and colletoketol (**4.83**). Col-letoketol (also known as grahamimycin) was also obtained from a *Cytospora* species. The hydroxyl-acids 5-hydroxyhex-2-enoic acid and 7-hydroxyoct-2-enoic acid, related to the tri- and tetraketide components of the macrodiolides, have been isolated from *C. capsici*. An interesting pair of 13-membered macrodiolides, bartanol (**4.84**) and an isomer bartallol, were also isolated. Their formation (Scheme 4.10) may arise by ring contraction of an epoxide (**4.86**→**4.87**), which is derived from the triene **4.85**. The epoxide is also an intermediate in the bio-synthesis of colletodiol.

## 4.8  Statins

The statins have achieved major importance as inhibitors of cholesterol bio-synthesis and they are used in humans for the treatment of coronary disease. These compounds block isoprenoid formation by inhibiting the key enzyme, hydroxymethylglutaryl co-enzyme A reductase (HMGCoA reductase). This pathway is described in Chapter 5. The active form of the statins, in which the lactone ring is opened, has a formal resemblance to HMGCoA. This inhibition has provided an effective way of lowering plasma cholesterol levels and in particular that of low density lipoprotein. Three of the statins in current clinical use (lovastatin, pravastatin and simvastatin) are fungal metabolites or derivatives of them. Lovastatin (mevinolin, monacolin K) (**4.88**) is produced by *Aspergillus terreus*, *Monascus ruber* and various *Penicillium* species. Compactin (**4.90**) is produced by *P. citrinum* and *P. brevi-compactum*. Pravastatin (**4.91**) is a microbiological hydroxylation product of compactin whilst simvastatin (**4.89**) is the 2,2-dimethylbutyryl ester related to lovastatin. Solistatin (**4.92**) is an aromatic compactin analogue that has been isolated from *P. solitum*, a fungus obtained as a contaminant of various cheeses.

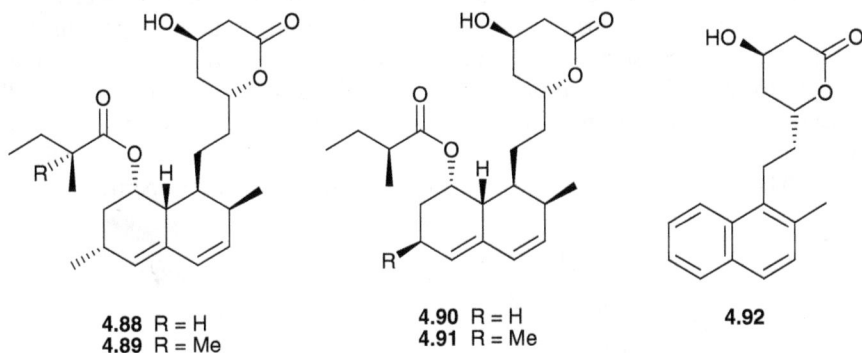

**4.88** R = H
**4.89** R = Me

**4.90** R = H
**4.91** R = Me

**4.92**

Biosynthetic experiments with variously labelled acetate units and [methyl-$^{13}$C] methionine established that the carbon skeleton of mevinolin was formed from nine acetate units, with the extra methyl group arising from methionine. The ester arose from two acetate units and a methionine. The oxygen atoms of the side chain at C-11, C-13 and C-15 were derived from the original acetate units whilst that at C-8 came from oxygen of the air. The cyclization reaction to form the bicyclic carbon skeleton may be a Diels–Alder cyclization (**4.93**→**4.94**). These are biosynthetically rather unusual. In the late stages of the biosynthesis, 4α,5-dihydromonacolin L (**4.95**) is hydroxylated and dehydrogenated to form mevinolin.

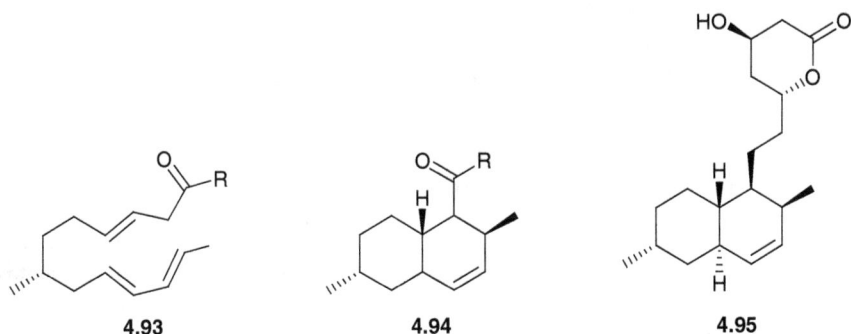

**4.93**

**4.94**

**4.95**

# 4.9 Cytochalasins

The cytochalasins are a quite widespread group of fungal metabolites. They are formed from an amino acid, typically phenylalanine, tryptophan or leucine, and an octa- or nonaketide. They have a characteristic structure exemplified by that of cytochalasin A (**4.96**), which was isolated from *Helminthosporium dematioideum*. There are several related compounds, such as the chaetoglobosins, exemplified by chaetoglobosin A (**4.97**) isolated from *Chaetomium globosum*, the aspochalasins isolated from *Aspergillus microcysticus*, the zygosporins isolated from *Zygosporim mansonii* and the phomins from a *Phoma* species. Many of these metabolites are responsible for the phytotoxicity of their parent organisms. These compounds have attracted considerable biological interest because they can bind to actin filaments and block its polymerization. Actin is a globular protein that provides mechanical support for the cell. A consequence of this binding includes changes to the cellular morphology and the inhibition of processes such as cell division. The cytochalasins affect cytokinases but not nuclear division. In mammalian cells these effects have also been observed on the aggregation of microtubuli. The cytochalasins also possess antibiotic and antiviral properties.

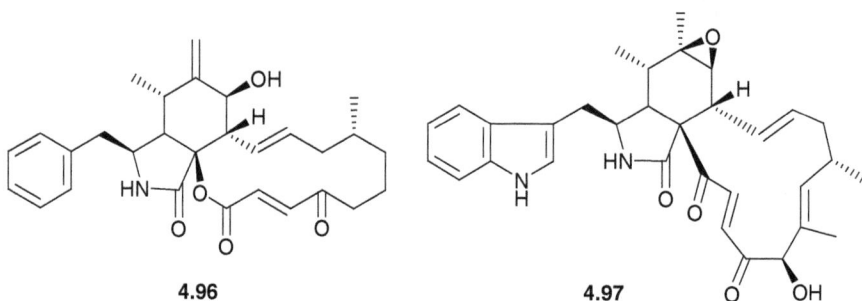

4.96                                            4.97

Labelling studies have shown that chaetoglobosin is assembled from tryptophan (**4.98**) and a nonaketide with three extra methyl groups that are introduced from methionine (Scheme 4.11). The polyketide chain is probably attached as an amide and it cyclizes to a tetramic acid. Treatment of the fungus with cytochrome $P_{450}$ inhibitors led to the accumulation of less highly oxidized metabolites, including prochaetoglobosin I. The six-membered ring of this intermediate is formed by an enzymatic Diels–Alder reaction based on the polyene **4.99** → **4.100**. The oxygen of the lactone ring of the cytochalasins arises by a biological equivalent of the Baeyer–Villiger reaction.

# 4.10 Fatty Acids from Fungi

Fatty acids play several important roles in fungi, in which they occur as the free acids and as esters, particularly of glycerol. They form energy storage metabolites whilst the phospholipids are components of membranes. The common fatty acids are oleic acid, linoleic acid, palmitic and stearic acids although

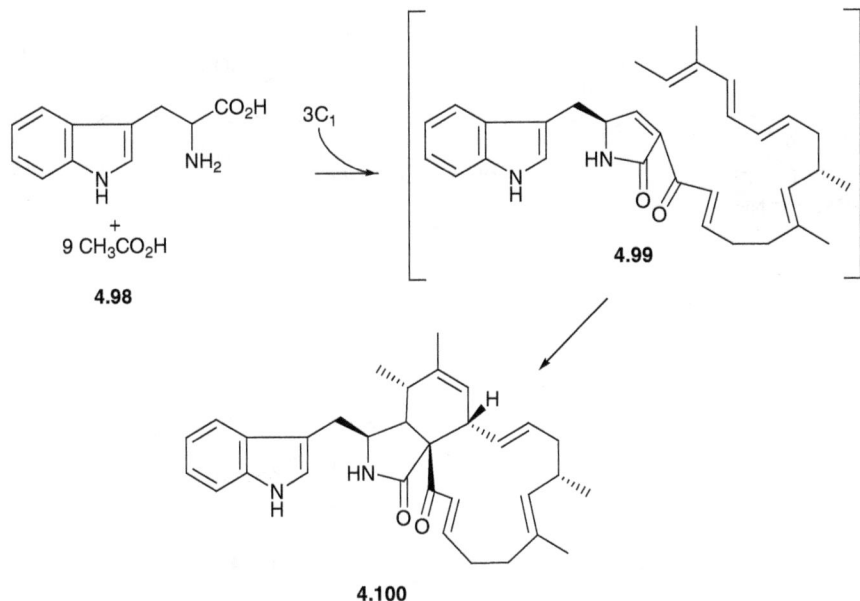

**Scheme 4.11**   Biosynthesis of the cytochalasins.

several other fatty acids have been detected. Individual fungi have a distinctive fatty acid profile that can be used to identify them. Some organisms, for example *Oospora lactis*, produce relatively large quantities of fats from carbohydrates and have been considered as a commercial source of fat. The production of polyunsaturated fatty acids by *Mortierella alpina* and *M. isabellina* has been examined in detail. These acids include the $C_{18}$ acids linoleic acid and γ-linolenic acid as well as the $C_{20}$ acids eicosapentaenoic acid and arachidonic acid.

Although there is a formal similarity to the biosynthesis of the polyketides, there are significant differences in the stereochemistry of fatty acid synthase. In the reduction of the enoyl thiol ester to generate the methylenes of the fatty acid, the acetate-derived hydrogen occupies the (*R*)-configuration. In growing polyketide chains, this hydrogen occupies the opposite (*S*)-configuration. Oleic acid is formed by dehydrogenation of stearic acid. The *cis* 9,10-double bond is formed by the loss of the C-9 pro-*R* and C-10 pro-*R* hydrogen atoms. Various modifications occur to the underlying fatty acid carbon skeleton, including ω- and ω-1 hydroxylation and lipoxygenase oxidation in the middle of the chain. Some of the volatile fungal metabolites that contribute to fungal odours are based on oct-3-en-1-ol, which arises from fatty acids.

The fungus *Gibberella fujikuroi* produces the gibberellins and the pigment bikhaverin. Some strains also produce the fusarins, *e.g.* **4.101**, and fumanosins, *e.g.* **4.102**, which are serious mycotoxins and phytotoxins. Cyclopentane fatty acids of the jasmonic acid type such as *N*-jasmonyl isoleucine (**4.103**) and its dihydro derivative have also been isolated from *Gibberella fujikuroi*.

**4.101**

**4.102**

**4.103**                                          **4.104**

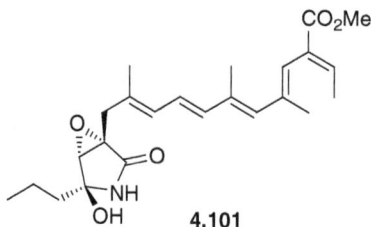

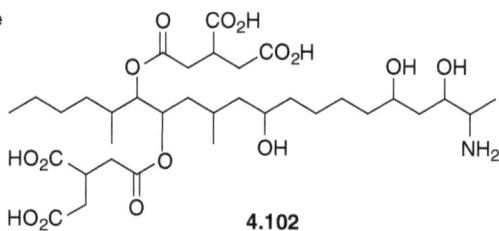

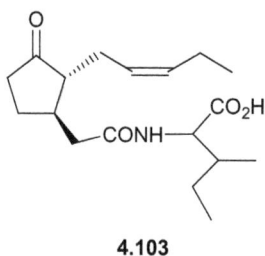

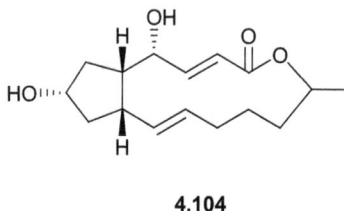

Brefeldins A (**4.104**) and B, obtained from *Penicillium brefeldianum* and which also possess a cyclopentane ring, have a wide spectrum of biological activity, including antiviral and cytotoxic effects. Brefeldin A has been shown to incorporate acetate and specifically [9-$^{14}$C]palmitic acid, suggesting that it is biosynthesized by a fatty acid pathway. However, other macrolactones such as curvularin appear to be biosynthesized by a polyketide pathway.

## 4.11 Polyacetylenes from the Higher Fungi

Compounds containing chains of conjugated triple bonds are widespread in the Basidiomycetes. They were first reported in 1952. For the most part they have been obtained from the culture fluid supporting a mycelial growth of the fungus, although a few have been detected in fruiting bodies. Their presence is relatively easily detected by their characteristic ultraviolet absorption although in the extraction of the fruiting bodies their presence may be masked by large quantities of ergosterol. However, their isolation in the pure state is more difficult because of their thermal instability. Indeed, some crystalline samples have been reported to explode when their melting point was being determined. A systematic screening of some 300 different Basidiomycetes showed that at least 10% produced readily detectable amounts of polyacetylenes in their culture fluids.

Several polyacetylenes have antibiotic properties and some are phytotoxic. Agrocybin (**4.105**), originally isolated from *Agrocybe dura*, is also formed by the 'fairy-ring' fungus, *Marasmius oreades*, and it may be responsible for the death of grass. Their production as extracellular mycelial metabolites may thus facilitate the spread of the mycelium in a hostile environment. The dry-rot fungus, *Serpula lacrymans* (*Merulius lacrymans*), is a common fungus found on wood in buildings where the wood has been damaged by a leaking roof or pipe. The hyphae penetrate the wood, producing lignases that break down the wood. As the wood dries, it cracks into characteristic small blocks. The hyphae may aggregate into mycelial cords that can then spread across adjacent brickwork and from which the fruiting bodies appear. The polyacetylenes that are formed are typified by **4.106**. These can reach a concentration of $30\,mg\,L^{-1}$ of the culture broth after 50 days growth. Polyacetylenes have been detected as trace constituents of fruiting bodies. *Fistulina hepatica* (the beef-steak fungus) produces the tetraynetetraol (**4.107**) as the major constituent of the broth but only small amounts of acids were detected in the sporophores. The cinnatriacetins (**4.108**) are triyne antibacterial agents that are also produced by this fungus.

**4.105**

**4.106**

**4.107**

**4.108**

There are several curious features about the biosynthesis of the compounds. They are formed by the fatty acid pathway involving the condensation of acetate and malonate units. Although an alkyne is at the same oxidation level as a methyleneketone, the alkyne is generated by reduction of the methylene-ketone to two methylenes and subsequent dehydrogenation to an alkene and thence an alkyne. The position of the alkyne in the chain does not necessarily correspond to that of a single acetate : malonate unit. Although the carbon chain of the polyacetylene may be only eight, nine or ten carbon atoms long, the acetate : malonate units are linked together to form a $C_{18}$ fatty acid that is then degraded. The pathway that has been established involved conversion of oleic acid *via* linoleic acid into crepenynic acid (**4.109**). The carbon chain is then cleaved to afford the $C_{10}$ and $C_8$ polyacetylenes such as dehydromatricaria acid (**4.110**) and its methyl ester. Interestingly, the eight-carbon chain is derived from carbon atoms C-9–C-16 of the chain with the loss of two carbons from the distal end of the chain. Polyacetylenes containing an odd number of carbon atoms are formed by elimination of the C-18 carbon atom by decarboxylation.

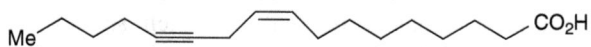

**4.109**

**4.110**

CHAPTER 5
# Terpenoid Fungal Metabolites

## 5.1  Introduction

Terpenoids are widespread natural products that are formed from $C_5$ isoprene units leading to their characteristic branched chain structure. Terpenoids are divided into families on the basis of the number of isoprene units from which they are formed. Thus there are monoterpenoids ($C_{10}$), sesquiterpenoids ($C_{15}$) diterpenoids ($C_{20}$), sesterterpenoids ($C_{25}$), triterpenoids ($C_{30}$) and carotenoids ($C_{40}$). The isoprene units are normally linked together in a head-to-tail manner. However, the $C_{30}$ triterpenoids and $C_{40}$ carotenoids are formed by the dimerization of two $C_{15}$ and $C_{20}$ units, respectively. Hence, in these cases the central isoprene units are linked in a head-to-head manner. The presence of tertiary centres in the isoprenoid backbone of the terpenoids facilitates skeletal rearrangements in the biosynthesis of these natural products. As a consequence, on first inspection some structures appear not to obey the isoprene rule.

There are numerous terpenoid fungal metabolites. In addition, several metabolites contain isoprene units linked to a carbon skeleton that has been biosynthesized by another pathway. The lysergic acid moiety of the ergot alkaloids contains an isoprene unit attached to the indole of a tryptophan. Fungal metabolites known as meroterpenoids contain a terpenoid fragment, typically that of a sesquiterpenoid, attached to a polyketide. However, some fungal metabolites such as nectriapyrone, which have a $C_{10}$ apparently terpenoid structure, are in reality polyketides with extra methyl groups arising from methionine.

## 5.2  Biosynthesis of Fungal Terpenoids

Isopentenyl diphosphate (pyrophosphate) (**5.5**) provides the isoprene unit of the terpenoids and steroids. There are two major routes by which it is formed (Scheme 5.1). The first involves mevalonic acid (**5.4**) and the second, quite

The Chemistry of Fungi
By James R. Hanson
© James R. Hanson, 2008

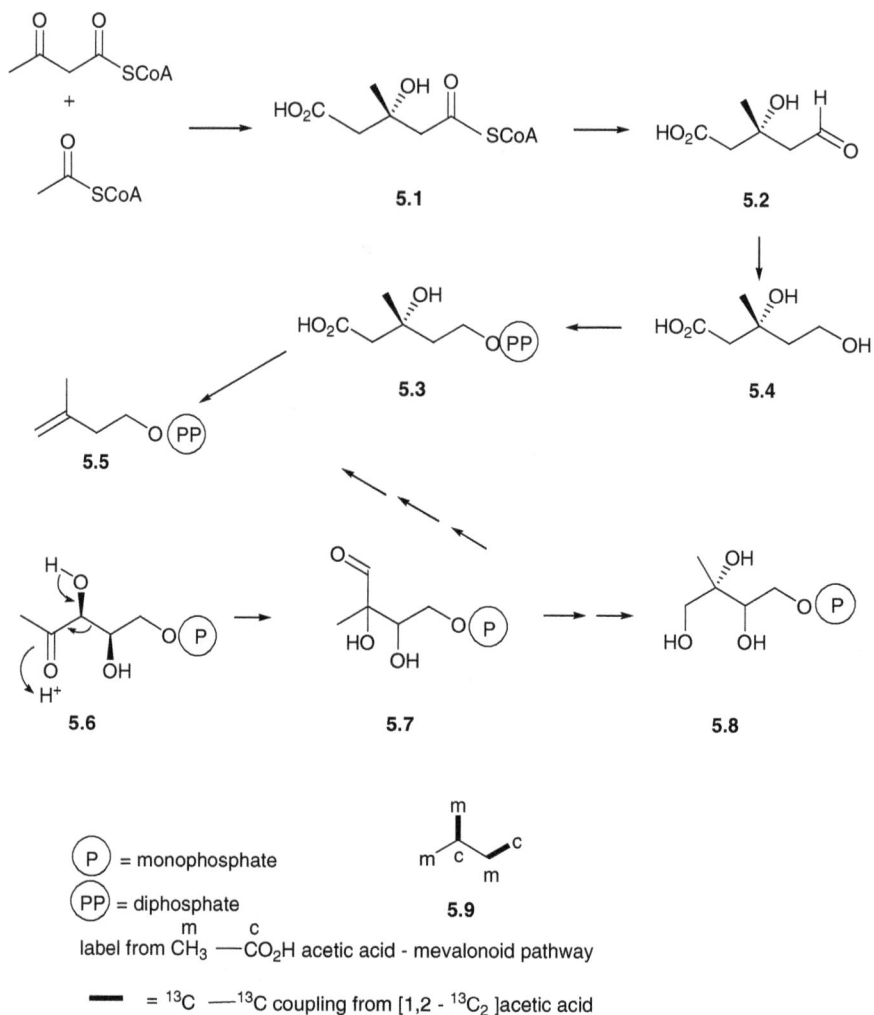

**Scheme 5.1** Biosynthesis of isopentenyl diphosphate.

different route, involves 1-deoxy-D-xylulose 5-phosphate (**5.8**) *via* **5.6** and **5.7**. Both pathways are found in higher plants and the non-mevalonate pathway predominates in the Streptomycetes. However, the mevalonate pathway, which starts from acetyl co-enzyme A, is the major route to isopentenyl diphosphate in fungi. The key stages in the mevalonate route involve the formation of hydroxymethylglutaryl co-enzyme A (**5.1**) and its reduction *via* the aldehyde (**5.2**) to (3*R*)-mevalonic acid (**5.4**). This is then converted into its diphosphate **5.3**, which undergoes decarboxylation to form isopentenyl diphosphate (**5.5**). (3*R*)-Mevalonic acid contains three prochiral centres, C-2, C-4 and C-5. Each of these is involved in the biosynthesis of the terpenes. Samples of mevalonic acid have been prepared in which each of these centres has been

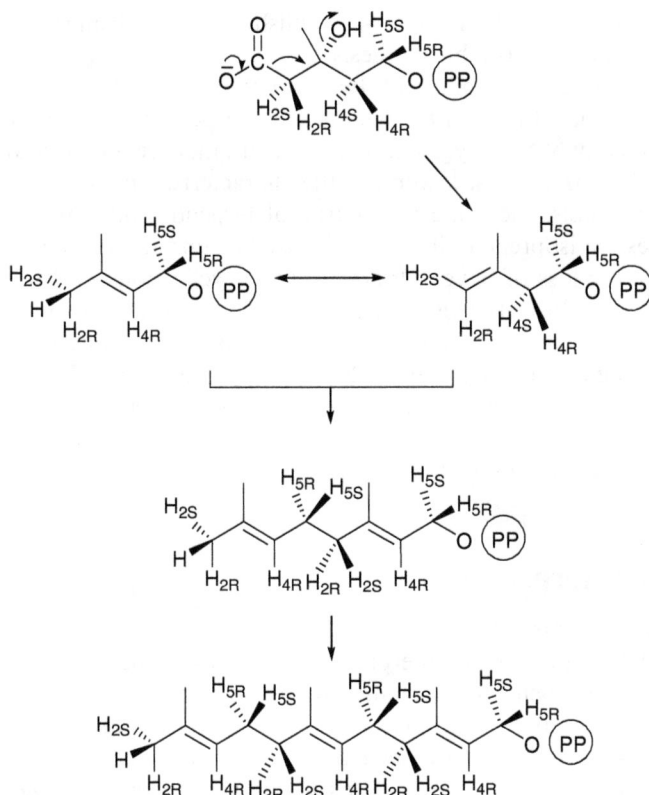

**Scheme 5.2**   Stereochemistry of the biosynthesis of farnesyl diphosphate.

stereospecifically labelled with deuterium or tritium. These labelled samples have been used to elucidate the stereochemistry of many of the major steps in the biosynthesis of the terpenes and steroids (Scheme 5.2). Thus, the decarboxylation of mevalonic acid diphosphate to form isopentenyl diphosphate takes place with the stereochemistry shown in Scheme 5.2. The fate of the carbon atoms from acetate units and the $^{13}C$–$^{13}C$ coupling pattern from [1,2-$^{13}C_2$]acetate in the resultant isoprene units are also shown in **5.9**.

Oligomerization of the isoprene units is based on two aspects of the reactivity of isopentenyl diphosphate. The first is the susceptibility of the alkene to electrophilic attack and the second is the displacement of the diphosphate with the formation of an electrophilic carbon. The isomerization of isopentenyl diphosphate to dimethylallyl diphosphate generates an allylic diphosphate which facilitates this process. Scheme 5.2 sets out the formation of geranyl ($C_{10}$) and farnesyl ($C_{15}$) diphosphates and the stereochemical fate of the mevalonoid hydrogen atoms. Although these steps were originally established in the context of mammalian steroid biosynthesis, they have subsequently been identified in fungal systems. The ultimate fate of the mevalonoid hydrogen atoms in various fungal metabolites has been used to establish the stereochemistry of the later stages in fungal terpenoid biosynthesis.

The head-to-head coupling of two farnesyl diphosphate units to form squalene is discussed later in the section on triterpenoids (Section 5.7).

Whereas many of the oxygen atoms of the original acetate units are retained in polyketide biosynthesis, leading to the characteristic oxygenation on alternate carbon atoms, the same is not true of terpenoid biosynthesis. In these biosyntheses the isoprene units and the prenyl chains are formed by processes that involve the loss of oxygen functions. The cyclizations lead to predominantly, but not exclusively (*e.g.* triterpenoid biosynthesis), cyclic hydrocarbons. Ultimately, the majority of the oxygen functions are introduced by oxidative processes using oxygen of the air with a few steps involving a hydration. Most terpenoid biosyntheses can be divided into those stages involved in the formation of the parent ring system and those involved in the oxidative relationships between the individual terpenoids.

## 5.3   Monoterpenoids

Unlike higher plants, monoterpenoids are far less common as major fungal metabolites. In plants the monoterpenoids function as attractants for pollinating insects and as allelochemicals. Fungi are less reliant on volatile terpenoids to exert their ecological influence. However, some fungi do rely on insect vectors such as bark-boring beetles for their transmission. *Ceratocystis* species are the cause of serious plant diseases such as Dutch elm disease (*C. ulmi*), which is spread by a beetle, whilst *C. coerulescens* is a blue-staining fungus of wood. Unsurprisingly, monoterpenoids such as geraniol, nerol, linalool and α-terpineol are found as metabolites of *Ceratocystis* species. A few Basidiomycetes, particularly wild mushrooms, have also been reported to produce monoterpenes such as geraniol and several monoterpene hydrocarbons. There is the probability of insect vectors being involved in the transmission of their spores.

## 5.4   Sesquiterpenoids

Sesquiterpenoids are common as fungal metabolites and they are particularly widespread amongst the Basidiomycetes. Sesquiterpenoids may be formed by the cyclization of farnesyl diphosphate (**5.10**) by pathways that involve an interaction of the carbocation formed by loss of the diphosphate with either the central or distal double bond. A few sesquiterpenoids are also formed by a cyclization that involves protonation of the distal double bond. This pathway is reminiscent of that leading to the di- and triterpenes. The carbocations that are formed in these first stages may then undergo various further cyclizations and rearrangements to ultimately generate the parent sesquiterpenoid, often as a hydrocarbon. Subsequent oxidative steps lead to the individual sesquiterpenoids. We consider firstly those sesquiterpenoids whose carbon skeleton is generated by an initial reaction between the central double bond and a carbocation derived from far-nesyl diphosphate. We then discuss those sesquiterpenes that arise by the attack of the carbocation derived from farnesyl diphosphate on the distal double bond.

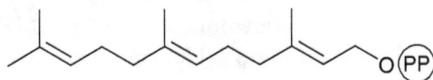

**5.10**

There can be different plausible foldings of farnesyl diphosphate that could generate sesquiterpenoid carbon skeleta. In many cases these have been differentiated by examining the labelling pattern of the sesquiterpenoid fungal metabolites that have been biosynthesized from variously labelled acetates and mevalonates. Recent interest has focussed on the isolation of sesquiterpenoid synthases that form these skeleta and the elucidation of the mechanism of their action.

## 5.4.1 Cyclonerodiol

The isomerization of farnesyl diphosphate (**5.10**) to nerolidol diphosphate (**5.11**) has been examined in a cell-free system from *Gibberella fujikuroi*. The cyclization of nerolidol diphosphate to cyclonerodiol (**5.12**) and the conversion of the latter into cyclonerotriol (**5.13**) have been observed in *Fusarium culmorum*.

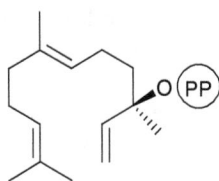

**5.11**

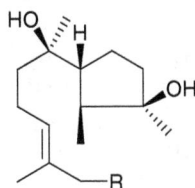

**5.12** R = H
**5.13** R = OH

Although bisabolenes are quite common plant constituents, they are relatively rare as fungal metabolites. Cheimonophylion A (**5.14**) is a highly oxidized bisabolene that was isolated from a Basidiomycete, *Cheimonophyllum candidissimium*, and which was toxic towards nematodes.

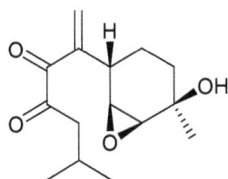

**5.14**

## 5.4.2   Helicobasidin

Helicobasidin (**5.15**) is a fungal pigment that was isolated from the plant pathogen *Helicobasidium mompa* in 1964. Its cuparene carbon skeleton has a similarity to that of the trichothecenes. Helicobasidin was shown to incorporate four tritium atoms from [2,2-$^3$H$_2$]mevalonate, two from (4$R$)-[4-$^3$H]mevalonate and two from [5,5-$^3$H$_2$]mevalonate. The incorporation of tritium from [2-$^3$H,2-$^{14}$C]geranyl diphosphate showed that the (4$R$)-mevalonoid hydrogen was retained from the second prenyl unit in the biosynthesis of helicobasidin. This led to the proposal of a hydrogen rearrangement during the biosynthesis similar to that found in the trichothecenes (see below). More recent studies of *H. mompa* have led to the isolation of some phenols that are probable precursors of the quinone.

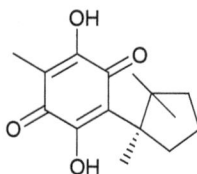

**5.15**

## 5.4.3   Trichothecenes

The trichothecenes are a major group of fungal sesquiterpenoids. The antagonism between the fungus *Trichothecium roseum* and various plant pathogenic fungi was first noted in 1909 and explored in the 1930s and 1940s. In 1947 Brian and Hemming showed that culture filtrates of *T. roseum* inhibited the germination of spores of the plant pathogen *Botrytis allii*. The anti-fungal agent trichothecin (**5.16**) was isolated by Freeman and Morrison from the cultures of *T. roseum* in 1948.

Trichothecin was shown to be the iso(cis)crotonic acid ester of a sesquiterpenoid, trichothecolone (**5.17**). Structural work reported in 1959 and 1960 showed that this hydrolysis product contained an unsaturated ketone and a secondary alcohol. Oxidation of the latter gave a cyclopentanone, trichothecadione. One of the remaining oxygen atoms was present in a reactive ether since hydrolysis of trichothecolone with hydrogen chloride gave a chlorohydrin (**5.18**) whilst dilute acid gave a glycol (**5.19**).

Although this reactivity is now known to be due to an epoxide, it was originally assigned to that of a four-membered oxetane. During this hydrolysis, both trichothecin and trichothecolone had undergone a rearrangement (*e.g.* **5.17** → **5.18**) that was not recognized at the time and led to a mis-interpretation of the degradative evidence. The key degradations (Scheme 5.3) that led to the identification of the carbon skeleton involved dehydrogenation reactions of various trichothecolone derivatives, *e.g.* **5.20** and **5.23**. These gave derivatives of *p*-xylene from one ring and 2,3-dimethylcyclopentenone from the other. Thus, tetrahydrothecadiol (**5.20**) gave *p*-xylene (**5.21**) and

**5.16** R = CCH=CHMe
          ‖
          O

**5.17** R = H

**5.18** R = Cl
**5.19** R = OH

**5.20**         **5.21**         **5.22**

**5.23**         **5.24**         **5.25**

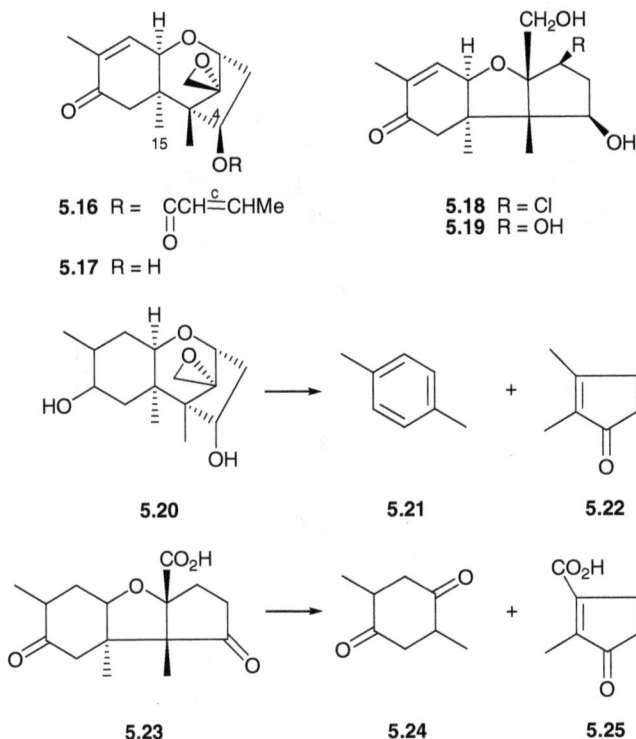

**Scheme 5.3**  Degradation of trichothecin.

2,3-dimethylcyclopent-2-enone (**5.22**) whilst **5.23** gave the 2,5-dimethylcyclo-hexa-1,4-dione (**5.24**) from one ring and the cyclopentanone acid **5.25** from the other. These fragments accounted for all 15 carbon atoms. Under basic conditions trichothecodione, which is a 1,5-diketone, undergoes a retro-Michael reaction, thus establishing the relationship between the carbonyl group on ring A and the hydroxyl group on ring C.

Although a structure for trichothecin was proposed in 1959, this proved to be incorrect. The correct structure was described in 1962 as a result of the identification of an epoxide ring in a relative, trichodermin (**5.26**), by $^1$H NMR and other studies. Trichodermin is an anti-fungal metabolite of a *Trichoderma* species and it was correlated with trichothecolone by oxidation and with another group of metabolites known as the verrucarols, which were obtained from *Myrothecium* species.

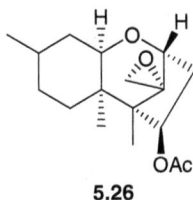

**5.26**

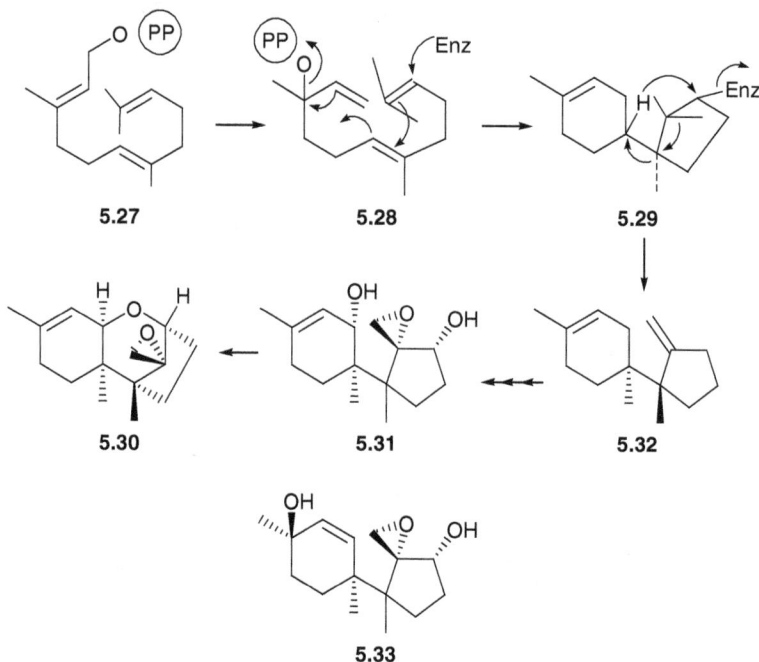

**Scheme 5.4**   Biosynthesis of the trichothecene skeleton.

Over 150 trichothecenes are now known, including about 80 simple non-macrocyclic fungal metabolites and 65 more complex metabolites that possess a trichothecene core and a macrocyclic ester bridging C-4 and C-15. They are produced by fungi from ten out of twelve sections of the genus *Fusarium* and by some members of taxonomically unrelated genera, including *Trichoderma*, *Trichothecium*, *Myrothecium* and *Stachybotrys*. Their biological activity as phytotoxins and mycotoxins is described in Chapters 8 and 9.

Most of the biosynthetic work has been carried out with trichothecin (Scheme 5.4). The sesquiterpenoid nature of trichothecin was established both by incorporation of three mevalonate units and of farnesyl diphosphate. NMR studies of the carbon-13 labelling pattern from [2-$^{13}$C]mevalonate and [1,2-$^{13}$C$_2$]acetate were consistent with the folding of farnesyl diphosphate as shown in **5.27**. Isomerization to nerolidyl diphosphate (**5.28**) preceded cyclization to the hydrocarbon trichodiene (**5.32**). The cyclization involved two methyl group rearrangements and a 1,4-hydrogen shift (**5.29**). The latter was elucidated by examining the fate of the [(4R)-4-$^3$H]mevalonoid hydrogen atoms. Thus, trichodiene retained all three [(4R)-4-$^3$H]mevalonoid labels even though the central isoprene unit possesses a quaternary carbon at the centre, originating from C-4 of mevalonate. Trichodiene synthase is one of the most thoroughly studied of the sesquiterpene cyclases. It has been shown that farnesyl diphosphate undergoes a rearrangement to nerolidyl diphosphate (an enzyme-bound intermediate). This allows the carbon chain to adopt a suitable geometry for cyclization. Analysis of the resultant trichodiene showed that the cyclization had proceeded with an overall retention of configuration at C-1 of the farnesyl diphosphate. Hydroxylation and epoxidation of trichodiene gave isotrichodiol (**5.31**), which in turn formed the parent trichothecene (**5.30**). At one time it was believed that trichodiol (**5.33**) was an intermediate but there is now evidence to suggest that this compound is an artefact. The stereochemistry of the mevalonoid hydrogen labels on ring C of the trichothecenes was established and used to show that the hydroxylations at C-3 and C-4 followed the normal patterns of retention of configuration.

### 5.4.4   PR-Toxin

The fungus *Penicillium roqueforti* is used in the production of blue-veined Roquefort, Stilton and Gorgonzola cheeses. It contributes to the flavour of these cheeses by degrading medium-chain fatty acids to methyl ketones with one less carbon atom. The fungus is also a microbial spoilage contaminant of dairy products and it is found on some mouldy grains. There are several sub-species of *P. roqueforti* that produce toxic metabolites which include the tremorgenic indoles and a group of sesquiterpenoids. PR Toxin (**5.34**) and the eremofortins are the best known of these highly oxygenated eremophilane sesquiterpenes.

The structure of PR-Toxin was established on the basis of the spectroscopic data of the parent metabolite and the changes to this data that arose as a result of simple reduction reactions. Its absolute configuration was established by anomalous dispersion in an X-ray structure. The toxicity has been associated with its inhibitory action on RNA and protein synthesis.

The labelling pattern of PR-Toxin biosynthesized from acetate and various mevalonates has been determined, revealing a hydrogen rearrangement of a $[(4R)-4-^3H]$mevalonoid hydrogen to create the secondary methyl group. The sesquiterpene hydrocarbon (+)-aristolochene (**5.35**) has been shown to be a key biosynthetic intermediate (Scheme 5.5). The enzyme system, aristolochene synthase, has been purified from *P. roqueforti*. It mediates the conversion of farnesyl diphosphate into the germacryl cation (**5.36**) and its subsequent rearrangement *via* the eudesmyl cation (**5.37**) into (+)-aristolochene (**5.35**). The crystal structure of the enzyme has been reported and the role of key amino acids in the terpenoid biosynthetic pocket in stabilizing the eudesmyl cation has been revealed by site-directed mutagenesis studies. The detection of (+)-aristolochene amongst the volatile metabolites of *P. roqueforti* has been used to identify those strains that might produce the PR Toxin.

### 5.4.5   Botryanes

The fungus *Botrytis cinerea*—a serious plant pathogen (Chapter 8)—produces a group of sesquiterpenoid fungal metabolites known as the botryanes. Botrydial (**5.38**), which is the major phytotoxin, and dihydrobotrydial (**5.39**) were first isolated in 1974. Their structures were established by X-ray crystallography.

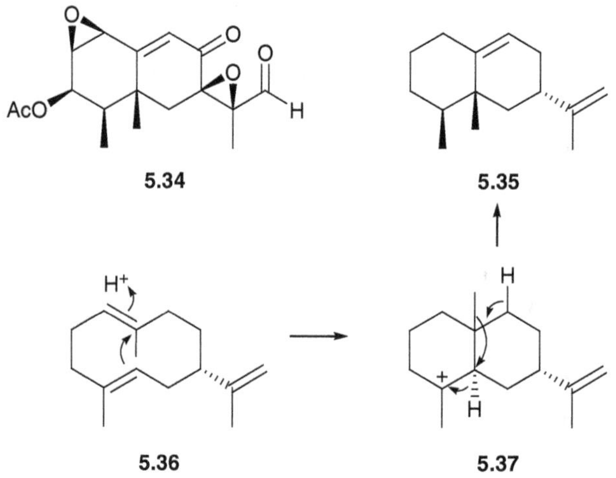

**5.34**                                                             **5.35**

**5.36**                                                             **5.37**

**Scheme 5.5**   Biosynthesis of aristolochene and PR-Toxin.

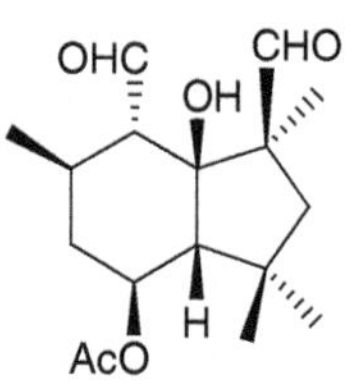

**5.38** 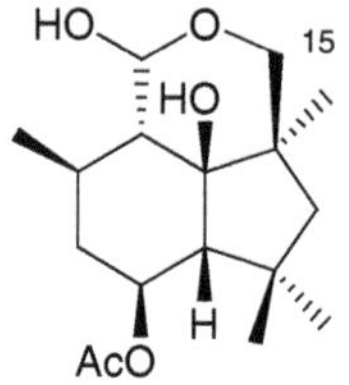                                                             **5.39**

At first sight these structures appear not to obey the isoprene rule. However, their sesquiterpenoid origin was established by the incorporation of farnesyl diphosphate. The botryane carbon skeleton could be formed (Scheme 5.6) by one of several different foldings of farnesyl diphosphate followed by a re-arrangement and bond cleavage (**5.40a–d**). The different plausible foldings of farnesyl diphosphate were distinguished by the labelling pattern arising from feeding [1,2-$^{13}$C$_2$]acetate,[4,5-$^{13}$C$_2$]mevalonate and the induced couplings arising from the cyclization and rearrangement of a multiply-labelled farnesyl diphosphate produced by a pulse-feeding experiment of [1-$^{13}$C]acetate to *B. cinerea*. In this experiment one of the plausible modes of cyclization and rearrangement of the multiply-labelled farnesyl diphosphate leads to bond formation and hence to $^{13}$C–$^{13}$C coupling between carbon atoms that had their origin in [1-$^{13}$C]acetate. The presence of these couplings in the final metabolite supported the folding **5.40d**.

The labelling pattern of the botryanes was consistent with the biosynthetic cyclization of farnesyl diphosphate (**5.41 → 5.42**). This was reminiscent of that leading to caryophyllene. The fate of the mevalonoid hydrogens shed further light on this process. All three [(4*R*)-4-$^3$H]mevalonoid hydrogens were in-corporated into dihydrobotrydial although one of the centres that would be labelled by C-4 of mevalonate is fully-substituted and carries a hydroxyl group

**5.40**

**5.41**   **5.42**   **5.43**

**5.38** ←

**5.44**   **5.45**   **5.46**

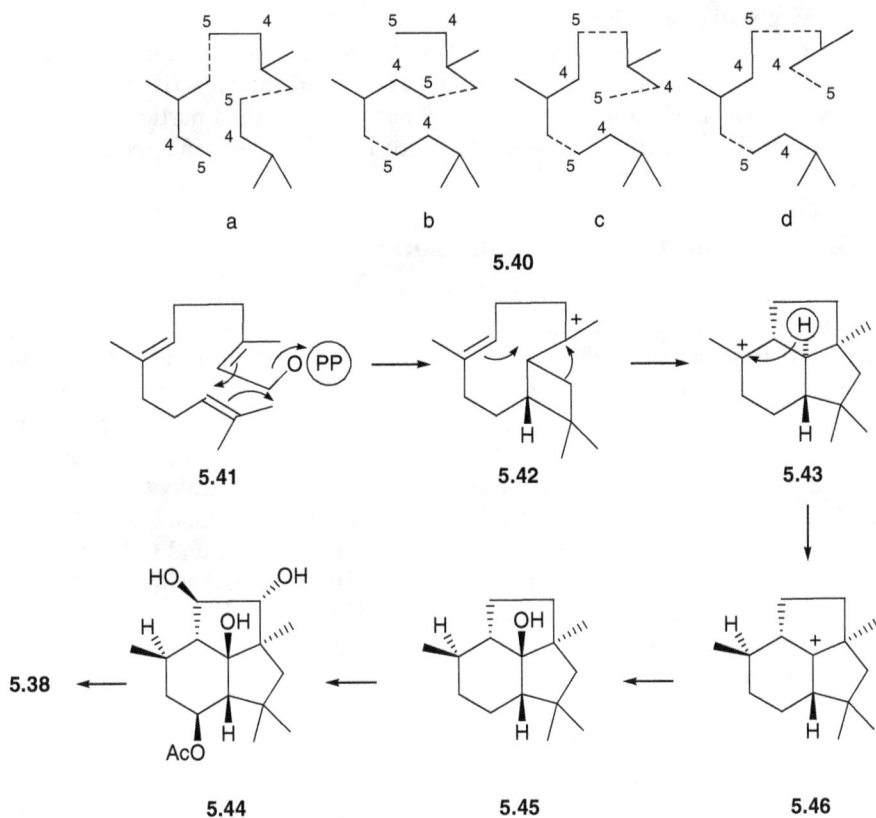

**Scheme 5.6** Biosynthesis of the botryane skeleton.

rather than a hydrogen atom. The biosynthetic scheme (Scheme 5.6) envisaged the rearrangement **5.43** → **5.46** of this mevalonoid hydrogen to generate a secondary methyl group. This rearrangement could take the form of two 1,2-shifts or a direct 1,3-shift. [4-$^2$H,4-$^{13}$C]Mevalonic acid was fed to *B. cinerea* at a concentration such that only one labelled mevalonate unit was incorporated into each botrydial that was biosynthesized. The retention of two $^2$H–$^{13}$C couplings established that a 1,3-hydrogen shift, as in **5.43**, had occurred since only one $^2$H–$^{13}$C coupling had been broken. The incorporation of oxygen-18 from [$^{18}$O]water revealed that the hydroxyl group at C-9 of **5.45** originated from water.

The oxidation level and stereochemistry of cleavage of the C-10–C-15 bond followed from several pieces of evidence. Dihydrobotrydial incorporated only one [2-$^3$H$_2$]mevalonoid label at C-15. Whereas botrydial was efficiently incorporated into dihydrobotrydial, the reverse transformation was significantly less efficient, suggesting that the dialdehyde was formed first. The incorporation of stereospecifically labelled [(2*R*)-2-$^3$H]- and [(5*R*)-5-$^3$H]mevalonates at C-15 and C-10, respectively, suggested that the cleavage might involve a *trans*

10β:15α-glycol (**5.44**). Subsequently, such a glycol was isolated. Further studies have led to the isolation of a series of probotryane and botryane metabolites which represent both precursors and metabolites of the bioactive metabolite, botrydial. Their relationship to botrydial has been explored particularly in the context of the control of the biosynthesis of this phytotoxic metabolite.

### 5.4.6   Culmorin and Helminthosporal

The sesquiterpenoid culmorin (**5.47**) was first described in 1937 by Raistrick during a study of the pigments of *Fusarium culmorum*. Since it was colourless, it was not investigated further until its structure was established by Barton in 1967. This structural evidence obtained during the 1960s reveals the impact that infrared and $^1$H NMR spectroscopy was making on structural, and in particular stereochemical, studies at that time. The strategy involved the sequential opening of the rings and degradation to the known $C_{10}$ tetrahydroeucarvone (**5.48**). The ent-longifolane carbon skeleton of culmorin was then established by reduction to ent-longiborneol. The configuration of the C-8 and C-11 alcohols was determined by analysis of the multiplicity of the CH(OH) $^1$H NMR signals, providing an early example of the power of these methods. Interestingly, the absolute configuration of the carbon skeleton of culmorin is the opposite to that of the plant product, longifolene.

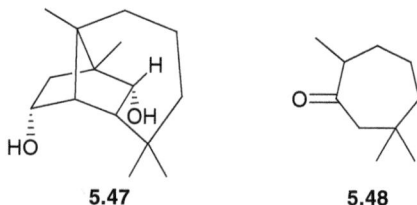

**5.47**                                    **5.48**

An array of oxygenated culmorin derivatives with additional hydroxyl groups at C-5, C-12, C-14 and C-15 as well as ketones have been isolated subsequently from *F. culmorum*, *F. graminearum* and *F. crookwellense*. The labelling pattern of culmorin when it is biosynthesized from labelled acetate and mevalonates is in accord with the folding of farnesyl diphosphate as shown in **5.49**.

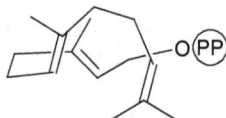

**5.49**

Helminthosporal (**5.50**) and helminthosporol (**5.51**) were first isolated from *Helminthosporium sativum*, now known as *Bipolaris sorokianum*. Their structures were established by de Mayo in 1962 and confirmed by synthesis by Corey in 1963. The structures were established at a time when $^1$H NMR spectroscopy was just beginning to make a contribution to structure elucidation and it is interesting to compare this with the work on culmorin (**5.47**), which was carried out a few years later. In the study of helminthosporal the $^1$H NMR spectrum was used to support the identification of the functional groups but not to interrelate them. Helminthosporal (**5.50**) was shown to contain two aldehydes, one of which was an α,β-unsaturated aldehyde. The relationship between these two functional groups was established by simple oxidation and reduction reactions, which gave a lactone and a dilactone. Dehydrogenation of the latter gave 6-isopropyl-3-methylphthalic anhydride (**5.52**), which contained 12 of the 15 carbon atoms of helminthosporal. The structure of this degradation product was established by synthesis, thus allowing a complete structure for helminthosporal to be proposed. However, later studies have shown that helminthosporal and helminthosporol are artefacts of the isolation process and are derived from acetals such as prehelminthosporol (**5.53**).

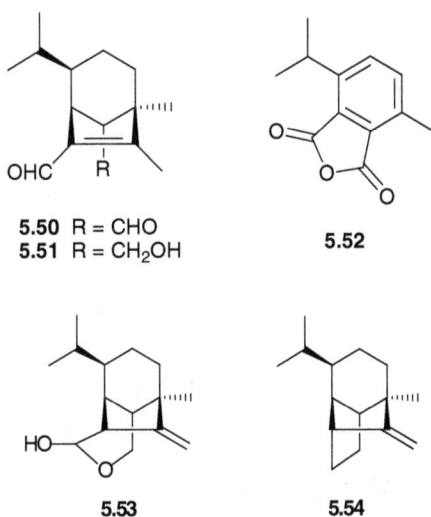

**5.50** R = CHO
**5.51** R = CH$_2$OH

**5.52**

**5.53**          **5.54**

A search amongst the hydrocarbons formed by *H. sativum* revealed the presence of the parent hydrocarbon sativene (**5.54**). The biosynthesis of these sesquiterpenes *via* this hydrocarbon has been examined.

## 5.4.7   Sesquiterpenoids of the Basidiomycetes

Sesquiterpenoids are widespread metabolites of the Basidiomycetes. *Lactarius* species (milk or ink caps) are common Basidiomycetes that are found growing in woodlands. They are characterized by a latex which is exuded when the fruiting body is cut or broken. The colour and taste of the latex is of taxonomic significance. The latex of the edible species, *Lactarius deliciosus*, is at first carrot-coloured. On exposure to air it then darkens and turns a green-blue colour. The colour has been associated with the formation of an azulene, lactaroviolin (**5.55**), which arises by the aerial oxidation of a dihydroazulene.

Lactaroviolin was first isolated in 1935. It was accompanied by an azulene, lactarazulene, which was shown to be dehydroguaiazulene, the dehydrogenation product of several plant sesquiterpenoids. The structure of lactaroviolin was established in 1954 by the work of Sorm. Wolff–Kishner reduction of the aldehyde gave lactarazulene, whilst the position of the aldehyde was located by a Grignard reaction with methylmagnesium iodide. The extra carbon atom derived from the Grignard reagent marked the position of the carbonyl group. The resultant carbinol was subjected to partial hydrogenation and hydrogenolysis to give the ethyl compound, which was identified by comparison with a synthetic sample. This Grignard strategy for locating the position of a carbonyl group in a natural product was quite widely used before physical methods became available. It is interesting contrast the way in which this structure was established with the spectroscopic methods that were used in 2006 to establish the structures of lactarazulenes, *e.g.* **5.56**, which were obtained from *Lactarius* species.

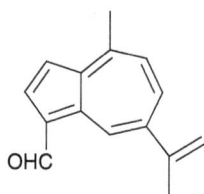 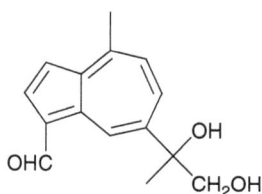

**5.55**                    **5.56**

The fruiting bodies of the higher fungi play an important role in the production and dissemination of their spores. The fruiting body must survive in a highly competitive environment by deterring microbial attack and that of insect and animal predators. Although many spores are distributed by air currents, others may require the intervention of insect or animal vectors. A large group of sesquiterpenoids play an important role in the chemical ecology of these interactions. Although their carbon skeleta are diverse a common feature is that many are derived (Scheme 5.7) from farnesyl diphosphate (**5.59**) by cyclization *via* humulene **5.60** and carbocation **5.61**. Rearrangement of the strained four-membered ring (pathway a) leads, *via* **5.58**, to the hirsutene skeleton (**5.57**). Alternatively (pathway b), rearrangement *via* the protoilludane (**5.64**) leads to the marasmane (**5.63**) and lactarane (**5.62**) or to the illudane (**5.65**) and sterpurane (**5.66**) carbon skeleta.

Although compounds with the monocyclic humulene carbon skeleton are relatively rare as fungal metabolites, the mitissimols, *e.g.* **5.67**, have been detected in *Lactarius mitissimus*. The fungal metabolite xenovulene A (**5.68**), which was obtained from *Acremonium strictum*, one of the Fungi Imperfecti, also contains a humulene ring attached to an unusual furocyclopentenone moiety. Labelling studies suggest that the latter may be derived from a C-methylated

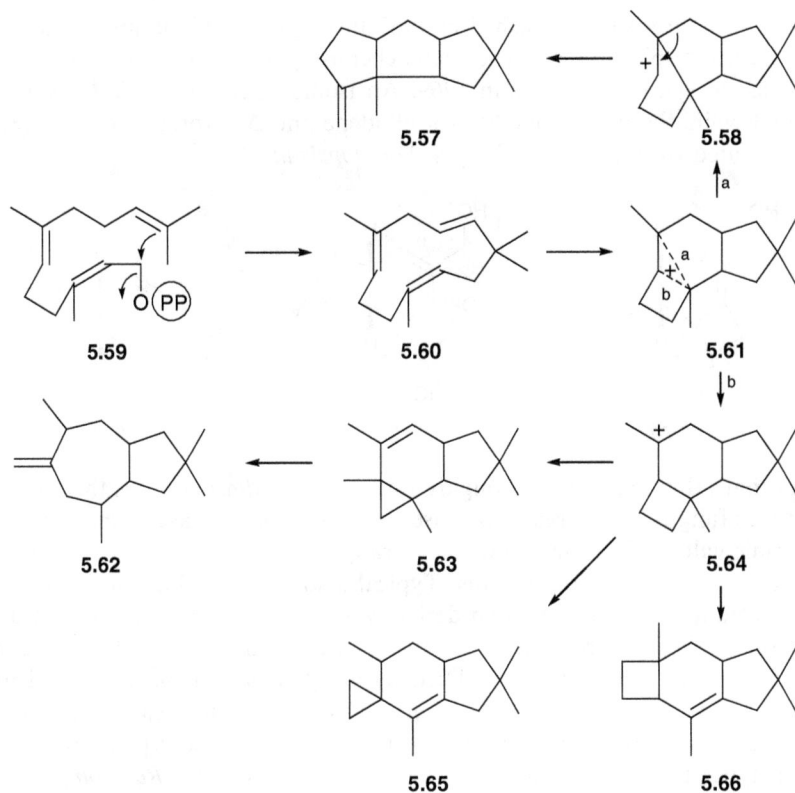

**Scheme 5.7** Biosynthetic relationships amongst the sesquiterpene skeleta derived from humulene.

polyketide. A few compounds with a *cis* caryophyllene skeleton such as hebelophyllene A (**5.69**), from *Hebeloma longicaudum*, are also known.

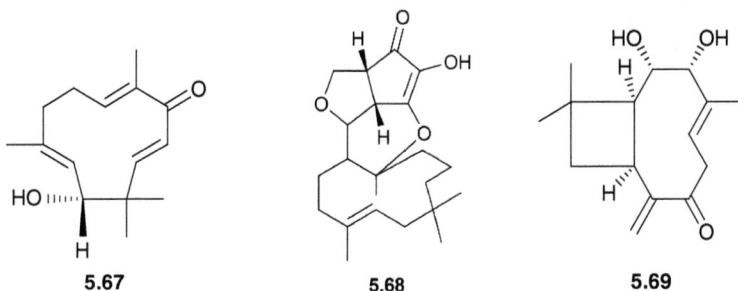

Compounds with the protoilludane carbon skeleton are widespread. The fungus *Clitocybe illudens* (*Omphalotus olearus*) is known as the Jack O'Lantern fungus because of its bioluminescence. Investigation of this fungus afforded both protoilludane and illudane metabolites. Illudol (**5.70**) was isolated in 1950 and its structure was established in 1967 through a combination of chemical and spectroscopic studies. The stereochemistry was established as a consequence of synthetic studies in 1969. An isomer, neoilludol (**5.71**), was also isolated while the precursors $\Delta^6$-protoilludene and $\Delta^7$-protoilluden-6-ol (**5.72**) were isolated from a bracket fungus, *Fomitopsis insularis*.

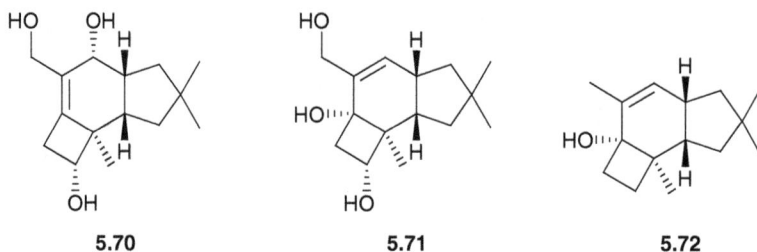

5.70                              5.71                              5.72

A fungus that has been thoroughly studied is *Armillaria mellea*, the honey or boot-lace fungus. This organism causes a serious root disease of many trees of economic value. The fungus produces a range of anti-bacterial and anti-fungal aromatic esters of protoilludenols. Typical acids are orsellinic acid and everninoic acid and their chlorinated derivatives. The structures of many of these compounds were established by Donnelly in the 1980s, primarily by spectroscopic methods. They are exemplified by armillyl orsellinate (**5.73**) and melleolide (**5.74**). Various other *Armillaria* species that also cause diseases of plants, *e.g. A. tabescens*, have been shown to produce sesquiterpene aroyl esters of this type. The dienone radulone A (**5.75**), obtained from *Radulomyces confluens*, is an interesting cytotoxic and anti-microbial member of this series. It is also a potent inhibitor of human platelet aggregation.

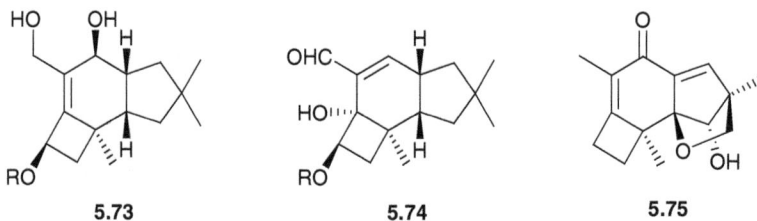

5.73                              5.74                              5.75

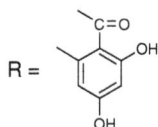

R =

The illudins M (**5.76**) and S (**5.77**) were isolated from *Clitocybe illudens* by Anchel and McMorris in 1950 and their structures were established in 1963. Illudin S was also known as lampterol, which was isolated from a Japanese bioluminescent fungus, *Lampteromyces japonicus*. Its structure was determined independently by a Japanese group in 1962. Many of the illudins, including those that have been isolated from other organisms, possess cytotoxic properties.

A particularly interesting anti-tumour isoilludane, HMAF (**5.78**), has been obtained from the treatment of illudin S with formaldehyde and sulfuric acid.

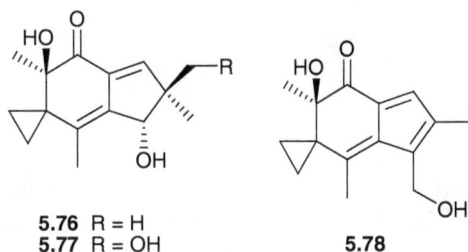

**5.76** R = H
**5.77** R = OH

**5.78**

The biosynthesis of the illudins from $^{13}$C-acetates and variously stereospecifically labelled mevalonates has been studied. The results are in accord with the cyclization of farnesyl diphosphate set out in **5.59→5.65**.

Another group of metabolites that have been isolated from *C. illudens* are the illudalanes. In these compounds the four-membered ring has been cleaved to give metabolites such as illudalic acid (**5.79**) and illudacetalic acid. Illudinine (**5.80**) is an alkaloid that can also be obtained from illudacetalic acid by treatment with ammonia.

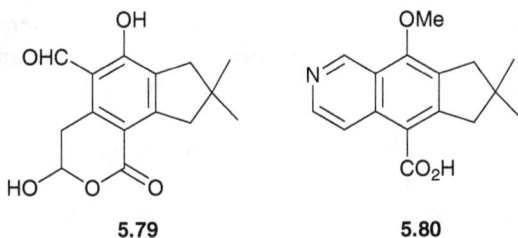

**5.79**

**5.80**

The fungus *Fomes annosus* is the cause of another serious root decay of trees, particularly conifers. The phytotoxic metabolite, fomannosin (**5.81**) also arises by cleavage of a protoilludane, this time of the six-membered ring. The cleavage is followed by a lactonization. Labelling studies with [1,2-$^{13}$C$_2$]acetate have confirmed this biosynthetic route.

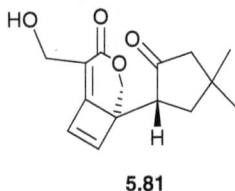

**5.81**

The 'silver-leaf' disease of fruit trees is caused by the fungus *Stereum purpureum*. The phytotoxic metabolites such as sterpuric acid (**5.82**) possess the sterpurane skeleton which may also be derived from a protoilludane by a series of rearrangements.

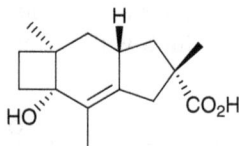

**5.82**

Marasmic acid (**5.83**)—first isolated from *Marasmius conigenus* in 1949—is a toxic metabolite that possesses anti-bacterial and mutagenic properties. Its structure was established in 1966 by a combination of chemical and spectroscopic studies and confirmed by X-ray analysis.

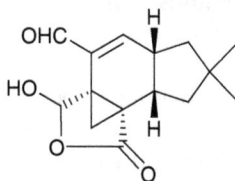

**5.83**

The marasmane carbon skeleton is quite widespread. Several *Lactarius* and *Russula* species have attracted interest because of their resistance to attack by insects, snails and small animals and because of their hot 'peppery' taste, which develops after they have been eaten. The constituents of the 'fleecy milk cap', *L. vellereus*, have been thoroughly studied. The primary sesquiterpenoid is the stearic acid ester of velutinal (**5.84**), which is stored in the lipid layer in the cell membrane. When the cell is damaged, lipases are released and the ester is converted rapidly into two toxic pungent dialdehydes, velleral (**5.85**) and iso-velleral (**5.86**). Isovelleral is mutagenic and both compounds have anti-microbial activity. These compounds are then gradually metabolized into less toxic metabolites. The picture is complicated by the fact that velutinal undergoes acid-catalyzed rearrangements to a series of furans. The combination of chemical and enzymatic reaction varies with different species and thus a range of marasmanes and lactaranes have been isolated, such as piperdial (**5.87**) from *L. piperatus* and the lactarorufins (*e.g.* A, **5.88**) from *L. rufus*.

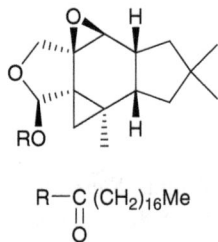

R—C (CH₂)₁₆Me

**5.84**

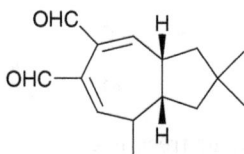

**5.85**

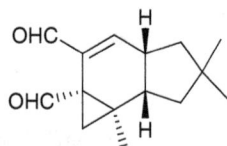

**5.86**

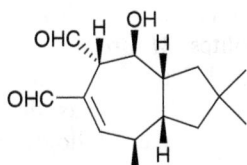

**5.87**

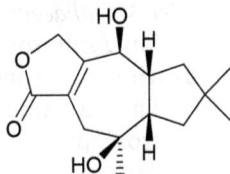

**5.88**

The anti-feedant activity of these compounds has been tested against various insects, including food storage pests such as *Tribolium confusum*.

The 'quaking' aspen tree (*Populus tremuloides*) is a common tree in North America that is used for its timber. The wood-rotting pathogen *Phellinus tremulae* causes significant economic losses of this tree. Although the sesquiterpenoid metabolites such as tremulenolide A (**5.89**) have a carbon skeleton that is isomeric with the lactaranes, some doubts have been expressed on its biosynthetic origin from farnesyl diphosphate.

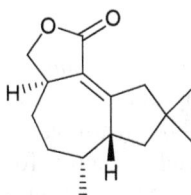

**5.89**

The hirsutane carbon skeleton, which is found in some other antibiotics produced by the Basidiomycetes, is also formed by a sequence of Wagner–Meerwein rearrangements of a protoilludane carbocation *via* **5.58**. Hirsutic acid (**5.90**) was isolated by Heatley in 1947 from *Stereum hirsutum* (*S. complicatum*) and its structure determined by X-ray crystallography by Scott in 1965. During the X-ray irradiation an unusual solid phase rearrangement occurred to give a carbonyl compound. The coriolins (*e.g.* A, **5.91**) with the same carbon skeleton were isolated from *Coriolus consors*. The isolation of the hydrocarbon hirsutene and the labelling patterns of hirsutic acid and the coriolins when they were biosynthesized from carbon-13 labelled acetates are in accord with their biosynthesis *via* a protoilludane.

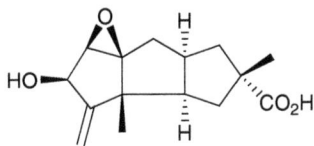

**5.90**

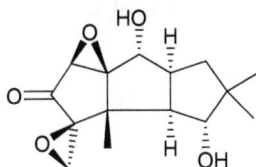

**5.91**

The genus *Marasmius* includes several small mushrooms that are found in both woodland and grassland. *Marasmius oreades* is the 'fairy ring' fungus

found in lawns whilst *M. alliaceus* is known as the 'garlic Marasmius' because
of the smell of its volatile sulfurous metabolites. *M. oreades* produces a group
of sesquiterpenoids with a drimane carbon skeleton exemplified by anhydromarasmane (**5.92**). The structure of this metabolite was determined by a
combination of spectroscopic methods and X-ray crystallography. The other
drimanes were correlated with anhydromarasmane. These metabolites have
plant growth regulatory activity. A polyacetylene, agrocybin (structure **4.105** in
Chapter 4), may be responsible for the phytotoxic effect of *M. oreades*.

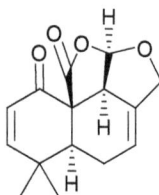

**5.92**

The sesquiterpenoid metabolites of *M. alliaceus* possess the alliacane carbon
skeleton and are exemplified by alliacol A (**5.93**) and alliacolide (**5.94**). This
unsaturated ketone inhibits DNA synthesis. The unsaturated lactone in the
structure may act as a Michael acceptor for biological nucleophiles. Biosynthetically, these compounds are interesting because there are several plausible ways by which this carbon skeleton may be derived from three isoprene
units. Whilst the substitution pattern of the six-membered ring is reminiscent of
a cadinane, the gem-dimethyl substitution of the five-membered ring suggests a
relationship to the protoilludanes. Thus there are two possible starting points
for incorporating the farnesyl chain, both of which require a rearrangement to
generate the alliacane skeleton of alliacolide. The labelling patterns arising from
the incorporation of $^2$H and $^{13}$C labelled acetate and mevalonate units were used
to define the isoprene units (**5.95**) and suggest that the metabolite might arise by
ring contraction of a cadinane (**5.96**). A cadinane, torreyol and the hydrocarbon
δ-cadinene have been isolated from the brown rot Basidiomycete *Lentinus
lepideus*. The bioluminescent fungus *Panellus stipticus* contains the cadinane
keto-aldehyde, panal (**5.97**). On treatment with ammonia or a primary amine in
the presence of iron(II) and hydrogen peroxide, light is emitted and thus panal
may be the precursor of the fungal luciferin in *Panellus stipticus*.

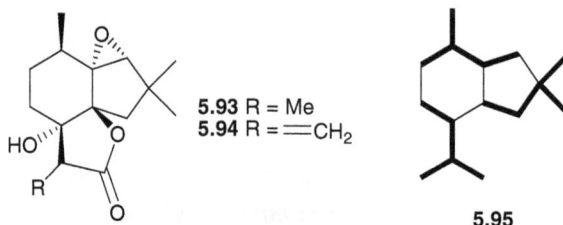

5.93 R = Me
5.94 R = =CH$_2$

**5.95**

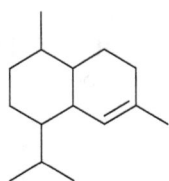

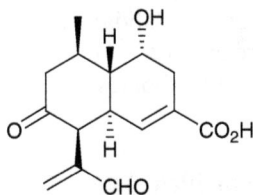

**5.96**                    **5.97**

## 5.5 Diterpenoid Fungal Metabolites

Fungal diterpenoids are formed by the cyclization of geranylgeranyl diphosphate. The cyclization may either involve an acid-catalysed polyene cyclization (Scheme 5.8) to form a bicyclic labdane (**5.98→5.99**). This may then be followed by a second cyclization initiated by the diphosphate acting as a leaving group to form tri- (**5.100**) or tetracyclic (**5.101**) metabolites. Alternatively, the initial cyclization may involve the diphosphate acting as a leaving group to generate a carbocation that alkylates the distal double bond in a biosynthetic sequence characteristic of the lower terpenes. Examples of the products of both cyclization sequences have been found amongst fungal metabolites. In those diterpenes that are formed by polyene cyclizations, the absolute stereochemistry of the A/B ring junction C-5 : C-10 may be either like that of the steroids (the so-called 'normal' stereochemistry) or it may be enantiomeric (the 'ent' stereochemistry). Examples of both series have been found amongst fungal metabolites.

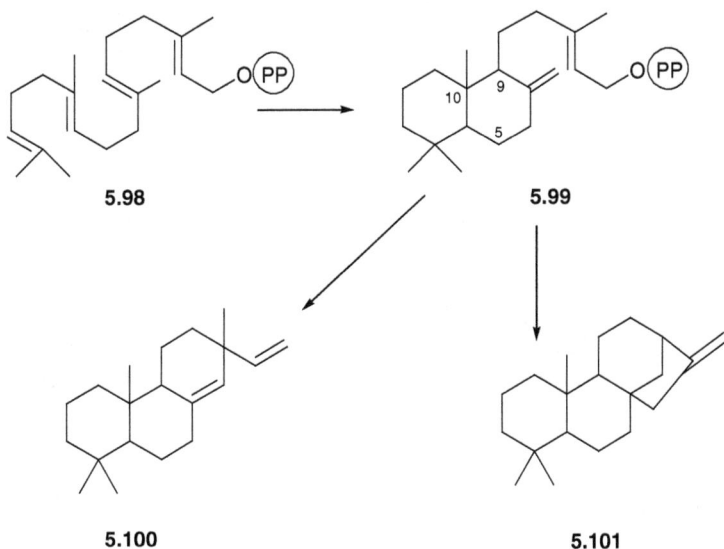

**5.98**          **5.99**

**5.100**                    **5.101**

**Scheme 5.8** Formation of the tri- and tetracyclic diterpenoid skeleta.

The first cyclization to a labdane may also generate a *cis* or a *trans* relationship between the hydrogen at C-9 and the methyl group at C-10. In this section we consider first those metabolites that arise by an initial polyene cyclization.

## 5.5.1  Virescenosides

The fungus *Oospora virescens* produces a family of tricyclic diterpenoid glycosides with anti-bacterial activity which is known as the virescenosides. The initial structural work was carried out with the virescenosides A (**5.102**) and B. The structures of the aglycones, virescenol A and B, were reported in 1968 in studies in which ¹H NMR spectroscopy played a supporting role to chemical degradation, ¹H NMR spectroscopy revealed that the acid-catalysed hydrolysis of the glycoside to release the aglycone also led to a double-bond isomerization from a tri- to a tetra-substituted position. Degradative and spectroscopic work revealed the presence of a 1,2-glycol, a primary alcohol, three tertiary methyl groups and a vinyl group in virescenol A. The primary alcohol and one of the secondary alcohols were removed by conversion into their toluene-*p*-sulfonates and hydrogenolysis with lithium aluminium hydride. When this was carried out with the double bond isomers, isovirescenols A and B, the product was the known 3β-sandaracopimara-8(9),15-diene (**5.103**). The position of the tri-substituted double bond was established by comparison of its ¹H NMR signal with those of related diterpenes. The structure of isovirescenol B has been confirmed by X-ray analysis.

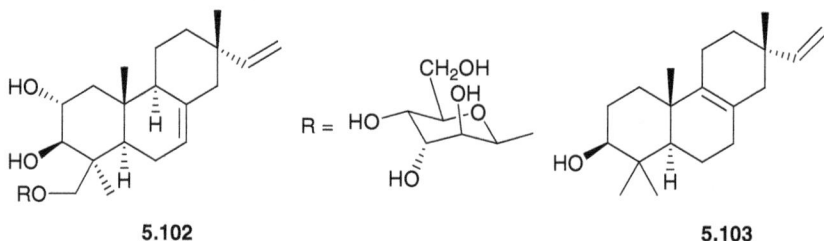

**5.102**                                                              **5.103**

The labelling pattern of isovirescenol A obtained by acid-catalysed hydrolysis of virescenoside A, which had been biosynthesized from [1-¹³C]-, [2-¹³C]- and [1,2-¹³C₂]acetates, was consistent with the diterpenoid origin of the metabolite and the intervention of a bicyclic copalyl diphosphate (**5.99**). The stereochemistry of the formation of ring C was revealed by the geometry of the deuterium labels on the vinyl group arising from the incorporation of stereospecifically labelled [5-²H]mevalonates. The biological $S_N2'$ displacement of the diphosphate follows an *anti* stereochemistry. The same stereochemistry was established for the cyclization leading to rosenonolactone, which is described below.

## 5.5.2  Rosanes

*Trichothecium roseum* produces a pink growth on decaying apples and it was investigated because it showed an antagonistic activity towards other fungi.

Examination of the culture broth afforded the sesquiterpenoid trichothecin, which has been discussed previously. Extraction of the mycelium gave the rosane diterpenoids, primarily rosenonolactone (**5.104**) and rosololactone (**5.105**). The elucidation of their structures in 1958 by Whalley came at a time when spectroscopic methods, in this case infrared spectroscopy, were beginning to have an impact on natural product chemistry. Thus the oxygen functions of rosenonolactone were shown to be a γ-lactone and a cyclohexanone by their IR absorption. The classical methods of structure determination in the terpenoid area involved degradation and structural simplification to give identifiable fragments. Thus the underlying carbon skeleton and the location of the carbonyl group was established by dehydrogenation to 1,7-dimethylphenanthrene and 9-hydroxy-1,7-dimethylphenanthrene. The basis of a key further degradation involved the ease of enolization of the ketone at C-7 to form a 7,8-enol. Oxidation of dihydrorosenonolactone with potassium permanganate gave a keto-acid that on treatment with alkali underwent a retro-aldol cleavage of the central ring to give two identifiable fragments, **5.106** and **5.107**, containing all the carbon atoms of rosenonolactone. The assignment of the stereochemistry of rosenonolactone made use of optical rotatory dispersion (ORD), which had been developed in the context of natural product structure determination in the 1950s by Djerassi and Klyne. Interestingly, the X-ray crystal structure of another rosane, rosein III (11β-hydroxyrosenonolactone) determined in 1970, was the first natural product structure to be established by non-heavy atom methods.

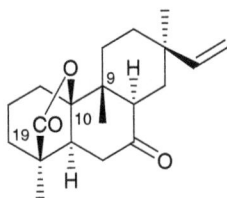

**5.104**

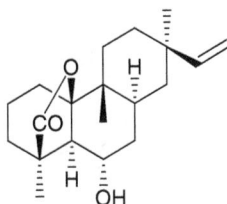

**5.105**

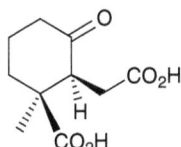

**5.106**

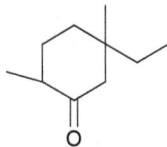

**5.107**

The biosynthesis of rosane diterpenoids has been thoroughly studied. The structure of rosenonolactone differs from that of the common tricarbocyclic diterpenes, in that the methyl group, normally at C-10, has migrated to C-9. However, the C-19:C-10 lactone ring is on the same face of the molecule as this methyl group, precluding a concerted rearrangement and lactonization.

The labelling pattern of rosenonolactone biosynthesized from [1-$^{14}$C]acetate and [2-$^{14}$C]mevalonate was reported independently by Birch and Arigoni in

1958. It established the diterpenoid origin of rosenonolactone. The formation of the tri- and tetracyclic diterpenoids from geranylgeranyl diphosphate takes place in two steps (Scheme 5.8). The first cyclization leads to the bicyclic copalyl diphosphate or its enantiomer whilst the second leads to the formation of the tri- and tetracyclic ring systems. Copalyl diphosphate (**5.108**) was incorporated into rosenonolactone. The stereochemistry of the allylic displacement of the diphosphate in the formation of ring C and the vinyl group was studied using deuterium NMR. The resonances of the protons of the vinyl group were assigned from their coupling patterns. The primary alcohol of the intermediate **5.108** was stereospecifically labelled by feeding (5$R$)- and (5$S$)-[5-$^2$H]mevalonic acids to *Trichothecium roseum*. The $^2$H NMR spectra of the resultant samples of rosenonolactone (**5.109**) enabled the stereochemistry of the deuterium atoms on the vinyl group to be established. This showed that the overall biological $S_N2'$ reaction involved in the displacement proceeds with an *anti*-stereochemistry as in the case of the virescenosides.

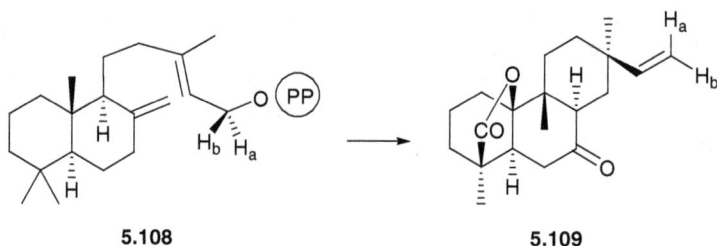

5.108                                                         5.109

The incorporation of (4$R$)-[4-$^3$H]mevalonic acid showed that the methyl group migration from C-10 to C-9 was accompanied by the shift of a hydrogen atom from C-9 to C-8. The number of 2-, 4- and 5-mevalonoid hydrogen atoms that were incorporated into rosenonolactone by *Trichothecium roseum* precluded alkene intermediates in the rearrangement. Furthermore, there was no loss of mevalonoid hydrogen from the migrating methyl group, excluding a cyclopropyl intermediate. Although pimara-8(9),15-diene (**5.110**) is formed by the cyclase, it is not an intermediate in the biosynthesis. A plausible explanation is that the rearrangement occurs with the formation of an enzyme-stabilized intermediate at C-10. Oxidation then takes place at C-19 and the formation of the lactone ring leads to the displacement of the substrate from the enzyme. The product, desoxyrosenonolactone (**5.111**), was shown to be converted into rosololactone (**5.105**) and, *via* the 7β-alcohol, to rosenonolactone (**5.104**).

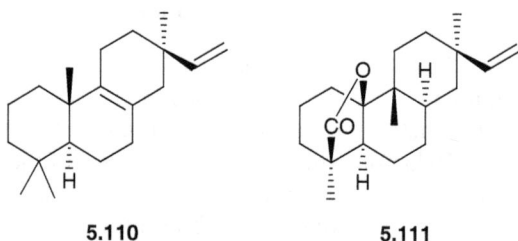

5.110                                                         5.111

### 5.5.3 Gibberellins and Kaurenolides

Gibberellins were originally isolated as the phytotoxic metabolites of a rice pathogen, *Gibberella fujikuroi*, in 1938. However, this material was impure and although some degradative work was reported in the Japanese literature during the 1940s it was confused by the lack of homogeneity of the natural material. The isolation of gibberellic acid (**5.112**) in 1954 enabled the structural work to proceed. There are now about 140 gibberellins that have been isolated from plants and fungi. The gibberellin plant hormones are widespread throughout the plant kingdom although they occur in very low concentrations. These plant hormones have the same carbon skeleton but differ in their oxygenation pattern. The fungus *Gibberella fujikuroi* is the main source of gibberellic acid, although gibberellins are also produced by five species of *Sphaceloma*, e.g. *S. manihoticola*, and also by *Neurospora crassa* and a *Phaeosphaeria* sp. L487.

The elucidation of the structure of gibberellic acid (**5.112**) reveals the impact of $^1$H NMR spectroscopy on structure determination. The evidence for the structure of gibberellic acid was obtained in the 1950s and early 1960s at a time of transition between the use of classical methods involving structural simplification and chemical degradation to identifiable fragments and the development of spectroscopic methods in which portions of the intact molecule were interlinked by their spectroscopic interactions. The underlying carbon skeleton of gibberellic acid was established (Scheme 5.9) by dehydrogenation to substituted fluorenes such as **5.113**. These were identified by comparison with synthetic samples. A key structural simplification of gibberellic acid involved the acid-catalysed conversion of ring A into an aromatic ring with the loss of a γ-lactone ring, one hydroxyl group and one isolated double bond. Allogibberic acid (**5.114**) was formed in the presence of dilute acid at room temperature whilst a rearrangement product, gibberic acid (**5.115**), was formed when the reaction was carried out at 100 °C. Apart from the Wagner–Meerwein rearrangement of rings C and D, this

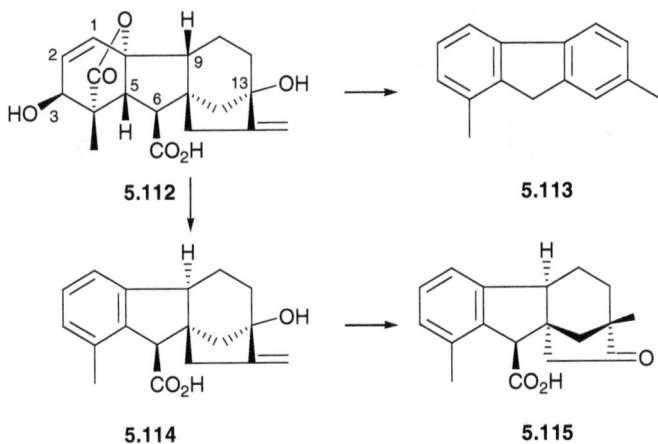

**Scheme 5.9**   Degradation of gibberellic acid.

degradation also involved an inversion of the stereochemistry at C-9. Stepwise degradation of allogibberic acid and gibberic acid led to the identification of the structures of rings B, C and D. The strategy involved, amongst other reaction sequences, opening ring D and selectively degrading ring C. The relationship of the functional groups on ring A of gibberellic acid then followed from oxidative reactions and the use of $^1$H NMR spectroscopy. The $^1$H NMR spectrum provided some evidence for the presence of the methyl group, the substitution pattern of the double bonds, the secondary alcohol and two key protons at C-5 and C-6. The allylic alcohol was oxidized to an α,β-unsaturated ketone. The stereochemistry of gibberellic acid was established by a combination of chemical and spectroscopic methods, including optical rotatory dispersion and X-ray crystallography.

Gibberellic acid undergoes a series of different rearrangements under both acidic and basic conditions. Although the interpretation of these rearrangements has provided considerable chemical interest, they were a source of confusion at the time of the structural work.

Examination of the culture broth of the wild type strain of *Gibberella fujikuroi* revealed the presence not only of further gibberellins such as the $C_{20}$ gibberellin, gibberellin $A_{13}$ (**5.116**), but also of a series of diterpenoids having the ent-kaurene carbon skeleton. The identification of the structures of these compounds paved the way for the study of the biosynthesis of gibberellic acid. The underlying carbon skeleton of the kaurenolides (**5.117**) was established by their conversion into the parent hydrocarbon, ent-kaurane. Ent-kaur-16-ene (**5.118**) had been obtained previously from the wood of New Zealand *Podocarpus* species. The position of the oxygen functions followed from their spectroscopic properties and the oxidative cleavage of the ring B diol which was partly masked by the lactone. Studies on the stereochemistry of the reactions of the kaurenolides played an important part in establishing the stereochemistry of ent-kaurene (**5.118**). An unusual aldehyde:anhydride, fujenal (**5.119**) in which ring B has been cleaved, has also been isolated from *G. fujikuroi*.

5.116

5.117

5.118

5.119

Two early observations on the biosynthesis of gibberellic acid were important. The labelling pattern of gibberellic acid biosynthesized from mevalonic acid established its diterpenoid origin. Secondly the hydrocarbon, ent-kaurene (**5.118**), was specifically incorporated into gibberellic acid, indicating the highly oxidative nature of its biosynthesis. Thus, there were three phases to the biosynthesis. The first involved the formation of ent-kaurene, the second involved the steps leading to the ring contraction to form the gibberellin carbon skeleton whilst the third stage involved the inter-relationship amongst the gibberellins, including the conversion of the $C_{20}$ into the $C_{19}$ compounds. Interestingly, it has now emerged that there are several differences between gibberellin biosynthesis in the fungus *Gibberella fujikuroi* and in higher plants.

Cyclization of geranylgeranyl diphosphate to ent-kaurene takes place in two stages. The first leads to ent-copalyl diphosphate and the second affords ent-kaurene. In plants there are two independent cyclases that mediate these steps and which are found in the plastids. In the fungus there is a bifunctional cyclase. The stepwise oxidation of ent-kaurene (Scheme 5.10) then takes place, firstly at C-19 to ent-kaurenoic acid (**5.120**). This is followed by hydroxylation at C-7 to form ent-7α-hydroxykaurenoic acid (**5.121**). This undergoes the ring contraction to form gibberellin $A_{12}$ 7-aldehyde (**5.122**). The labelling results with [2-$^{14}$C]mevalonic acid had shown that it was C-7 of ent-kaurene that was extruded to form the aldehyde. Studies with stereospecifically labelled ent-7α-hydroxykaurenoic acid showed that a C-6β hydrogen atom was lost in the ring contraction. However, the 6β,7β-diol was not a gibberellin biosynthetic intermediate although it was converted into the ring B seco-diterpenoid fujenal (**5.119**). The ring contraction is probably initiated by the oxidative abstraction of the hydrogen from C-6β.

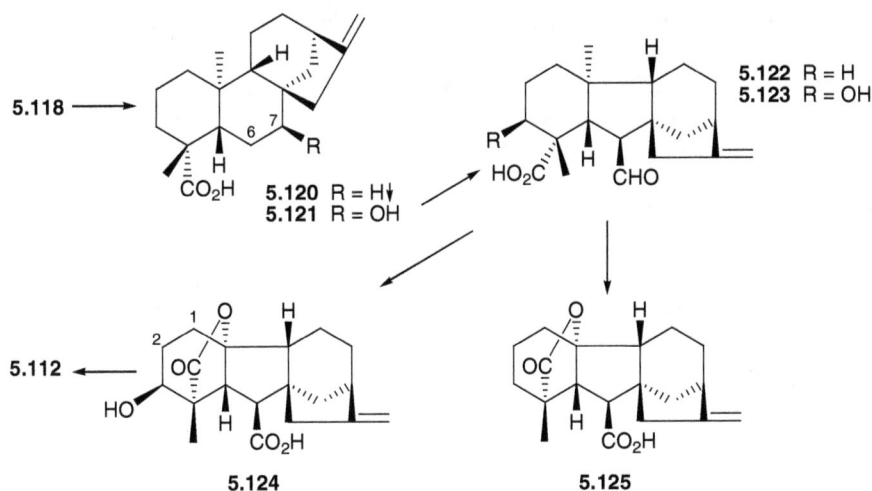

**Scheme 5.10** Biosynthesis of gibberellic acid.

The gibberellin biosynthetic pathways in *Gibberella fujikuroi* then diverge. In the main pathway hydroxylation takes place at C-3 to form gibberellin $A_{14}$ aldehyde (**5.123**) and thence gibberellin $A_{14}$. In fungi the oxidative removal of C-20 to form the $C_{19}$ gibberellins is mediated by a cytochrome $P_{450}$ mono-oxygenase whilst higher plants utilize an iron 2-oxoglutarate dioxygenase. Although C-20 is removed at the aldehyde oxidation level to form gibberellin $A_4$ (**5.124**), this carbon atom ultimately appears as $CO_2$. The C-20 carboxylic acids do not appear to be intermediates. Mevalonoid hydrogen is retained at C-1, C-5 and C-9, precluding alkene intermediates. Studies on the incorporation of mevalonoid hydrogen show that in the conversion of gibberellin $A_4$ into gibberellin $A_7$ the ring A double bond is formed by the *cis* loss of hydrogen from the $\alpha$-face of the molecule. Interestingly, gibberellins with 1$\alpha$- and 2$\alpha$-hydroxyl group ($GA_{16}$ and $GA_{47}$) are minor metabolites of *G. fujikuroi*. The final step in the fungal biosynthesis of gibberellic acid is hydroxylation at C-13. There is a minor pathway from gibberellin $A_{12}$ 7-aldehyde *via* the acid, gibberellin $A_{12}$ to gibberellin $A_9$ (**5.125**) that parallels the major pathway. In *Phaeosphaeria* species, oxidation of the gibberellin $A_{12}$ aldehyde leads to gibberellin $A_9$ and then to gibberellin $A_{20}$ and gibberellin $A_1$.

The kaurenolides and fujenal that are formed by *Gibberella fujikuroi* are side products from the major biosynthetic pathway. Ent-kaurenoic acid is dehydrogenated to form ent-kaur-6,16-dienoic acid. Epoxidation on the $\beta$-face and hydrolysis afforded 7-hydroxykaurenolide (**5.117**), which was then hydroxylated at C-3 or C-18 to form the dihydroxykaurenolides. The unusual aldehyde : anhydride fujenal (**5.119**) was formed *via* the 6$\beta$,7$\beta$-diol. Amongst the metabolites of *Gibberella fujikuroi* there are several compounds in which the 16-ene is hydrated to give a 16$\beta$-tertiary alcohol. These compounds are not metabolized further and represent a 'dumping mechanism'.

In common with a number of fungal secondary metabolites, the genes for gibberellin biosynthesis are clustered in *Gibberella fujikuroi*. The cluster includes genes that code for a geranylgeranyl diphosphate synthase, a bifunctional copalyl diphosphate/kaurene synthase and a group of cytochrome $P_{450}$ mono-oxygenases that mediate the oxidation of ent-kaurene to the 19-aldehyde, the further oxidation at C-19 and C-7 and the ring contraction. Other genes code for the oxidation at C-3 and the C-20 oxidase. Mutants lacking these genes have been identified. One of these, B-41a, proved particularly useful in work reported in 1975 that established important details of the biosynthetic pathway.

The distribution of the genes for gibberellin biosynthesis in members of the *Gibberella fujikuroi* species complex has been examined. Although the entire gene cluster is present in many members of this group, they are only expressed and lead to gibberellin biosynthesis in those strains that are rice pathogens. This observation is of general importance in the study of the phytopathogenicity of fungi.

Recently, a gibberellin receptor has been identified. The gibberellins mediate several plant developmental processes. They are associated with activities ranging from the induction of $\alpha$-amylase in the germination of seeds, to stem elongation to the development of flowers and fruit.

## 5.5.4 Aphidicolin

Aphidicolin is a diterpenoid metabolite of *Cephalosporium aphidicola* and *Phoma betae*. The structure of aphidicolin (**5.126**) was established in 1972 through a combination of chemical, spectroscopic and X-ray crystallographic studies. Unlike the ent-kaurene and gibberellin metabolites of *G. fujikuroi*, aphidicolin possesses an absolute stereochemistry that is 'steroid-like'. It has attracted considerable interest as a specific inhibitor of DNA polymerase α and as a potential anti-viral and anti-tumour agent.

The biosynthesis of aphidicolin has been examined (Scheme 5.11). The enrichment and coupling patterns of aphidicolin biosynthesized from [1-$^{13}$C]-, [2-$^{13}$C]- and [1,2-$^{13}$C$_2$]acetate were used to define the constituent isoprene units. The numbers of 2- and 5- mevalonoid hydrogens that were incorporated were established by tritium labelling. The generation of a $^2$H–$^{13}$C coupling, which was observed in the $^2$H NMR spectrum of aphidicolin biosynthesized from [4-$^2$H,3-$^{13}$C]mevalonic acid, established the rearrangement of a 9β-hydrogen atom to C-8 during the biosynthesis (**5.127**→**5.129**→**5.128**). The tertiary alcohol aphidicolan-16β-ol, formed by hydration of the cation **5.128**, was an efficient precursor of aphidicolin. It was shown that the tertiary C-16 hydroxyl group of aphidicolin arises from H$_2$$^{18}$O by hydration rather than by hydroxylation. Incorporation studies have suggested that hydroxylation then takes place at C-18 followed by C-3α and C-17. There is evidence for a minor pathway involving the 16-ene and 16β,17-epoxyaphidicolane-3α,18-diol.

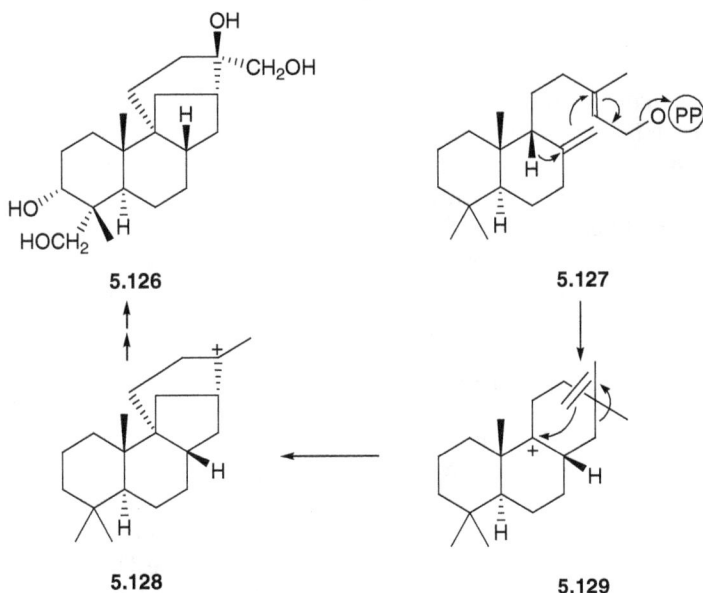

**Scheme 5.11**  Biosynthesis of aphidicolin.

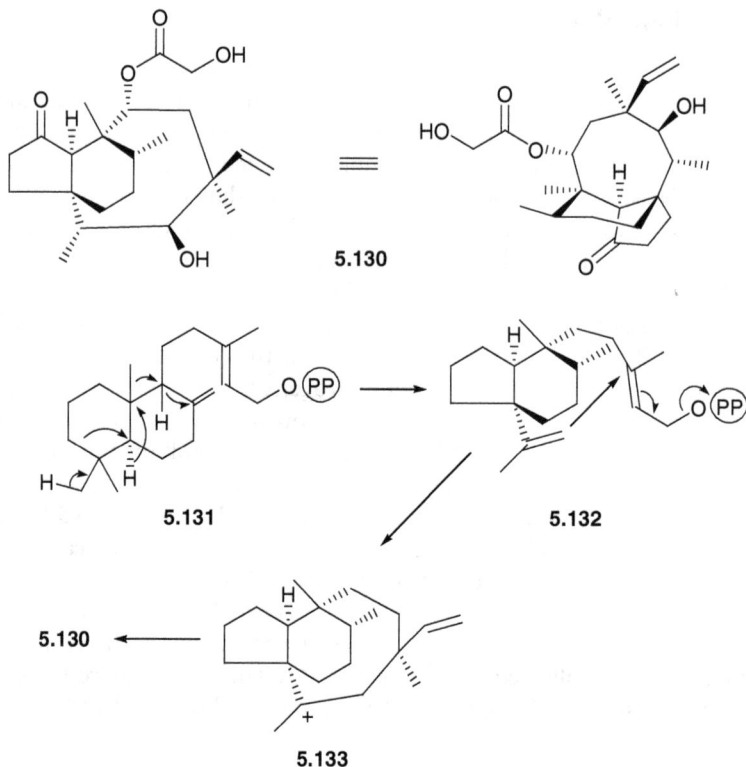

**Scheme 5.12**  Biosynthesis of pleuromutilin.

## 5.5.5  Pleuromutilin

The mevalonate labelling pattern of the antibiotic pleuromutilin (**5.130**), which is produced by the Basidiomycete *Pleurotus mutilus* (*Clitopilus scyphoides*), suggests that it is formed (Scheme 5.12) by a more deep-seated rearrangement of a copalyl diphosphate (**5.131**→**5.132**→**5.133**). A derivative, tiamulin®, is used in animal health and retapamulin® (altabax®) has attracted interest because of its activity against resistant organisms where it selectively targets the bacterial ribosome.

## 5.5.6  Fusicoccins and Cotylenins

The fungus *Fusicoccum* (*Phomopsis*) *amygdali* is responsible for a wilting disease of peach and almond trees. The principle phytotoxic metabolite is a glucoside, fusicoccin A (**5.134**). The structure of this diterpenoid fungal metabolite was established in 1968 by a combination of X-ray crystallographic and chemical studies carried out in Italy by Ballio and in the UK by Barton and Chain. The key degradations of the aglycone were based on the cleavage of the diol on its eight-membered ring and the formation of a cyclic ether between the

C-8 hydroxyl group and the C-2 of the ring B alkene. The mass spectroscopic fragmentation pattern and a careful assignment of the $^1$H NMR spectrum allowed the formulation of several part structures from which the final structure was deduced.

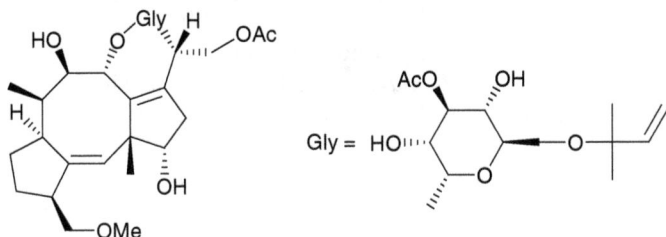

**5.134**

Several related metabolites have been identified, including fusicoccins H and J, which were shown to be precursors of fusicoccin A. Although there is a formal similarity between the fusicoccins and the sesterterpenoid ophiobolins, the fusicoccins are diterpenoids. They are closely related to the cotylenins, which are metabolites of *Cladosporium* species. The cotylenins have the same underlying diterpenoid carbon skeleton but differ in the modification of the glycoside unit. The labelling pattern of fusicoccin A biosynthesized from [1-$^{13}$C]- and [2-$^{13}$C]acetate and from [3-$^{13}$C]mevalonate is consistent with the folding of geranylgeranyl diphosphate, which is shown in **5.135**. The sites of incorporation of the deuterium labels from [4-$^2$H,3-$^{13}$C]mevalonate were established by their effect on the $^{13}$C NMR signals. This showed that the cyclization proceeded *via* a bicyclic intermediate and consecutive 1,2-hydride shifts. More recent studies have indicated that fusicocca-2,10(14)-diene (**5.136**) and the C-8β -alcohol are initial intermediates in the biosynthesis.

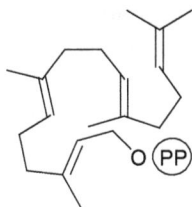

**5.135**

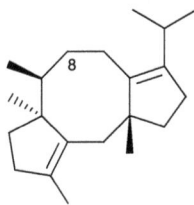

**5.136**

The biological activity of fusicoccin A appears to be associated with its effect on the plasma membrane, where it binds to a protein and interacts with the H$^+$-ATPase. The binding to the protein leads to an opening of the stomata, a consequent high loss of water and the wilting of the leaves. The interaction of the fusicoccins and coylenins with the plasma H$^+$-ATPase can also lead to some phytohormone-like effects.

# 5.6   Sesterterpenoids

Unlike the other families of terpenoids the $C_{25}$ sesterterpenoids are relatively rare and have only been studied during the last 40 years. They are mainly of fungal and marine origin. Only a few examples have been found in plants and insects. The ophiobolins are a group of sesterterpenoid fungal metabolites that are characterized by a tricyclic 5-8-5 structure (**5.137**). They are the phytotoxic metabolites of a group of plant pathogens that are now grouped within the genus *Bipolaris* as *B. oryzae*, *B. maydis* and *B. sorghicola*. The fungus *B. oryzae* was originally known as *Helminthosporium oryzae*, *Drechslera oryzae* or *Cochliobolus miyabeanus*. Ophiobolins have also been isolated from *Aspergillus ustus*. Some of these metabolites were studied independently by different groups who assigned them various trivial names. Thus, ophiobolin A (**5.137**) was also known as cochliobolin A or zizanin.

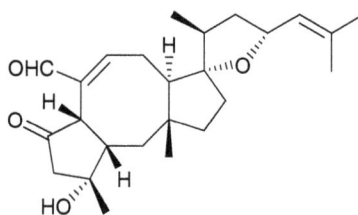

**5.137**

The organisms that produce these metabolites are serious pathogens of rice, maize and sorghum. The ophiobolins induce symptoms and physiological changes in the infected plants. They have several detrimental effects on plants, including reducing the growth of the roots and altering the permeability of cell membranes. There is an antagonistic interaction between these effects and some calmodulin-dependent processes.

The structure of ophiobolin A (**5.137**) was established by chemical degradation and by X-ray crystallography of a bromo derivative. The spectroscopic characteristics of ophiobolin A led to the identification of the oxygen functions and the double bonds. The relationship between the aldehyde and the cyclopentanone was revealed by the formation of a γ-lactone on reduction and partial re-oxidation. A cyclic pyridazine was formed with hydrazine. Tetrahydro-ophiobolin A formed an unusual cyclic peroxide involving these two groups. Vigorous oxidation of tetrahydro-ophiobolin A afforded a heptanoic acid lactone, which revealed the structure of the side-chain.

Ophiobolin B (**5.138**) was isolated from the culture filtrates of *Helminthosporium zizaniae* or *Ophiobolus heterostrophus*. It was related to ophiobolin A through a common hydrogenolysis product. Several other ophiobolins have also been isolated.

The ophiobolin carbon skeleton is formed by the cyclization of geranylfarnesyl diphosphate, as shown in **5.139**, although an isomerization of the 2-*trans* double bond to the *cis* isomer is required to generate the correct stereochemistry. The incorporation of [2-$^{14}$C]mevalonic acid and geranylfarnesyl diphosphate provided support for this scheme. A biosynthetic sequence linking ophiobolin C to ophiobolin B and thence ophiobolin A has been established. The location of the labels from the incorporation of $[2-^3H_2]$mevalonic acid revealed that there was a 1,5-hydride shift from C-8 to C-15 during the cyclization. The use of stereospecifically labelled mevalonates showed that this was the 8β-hydrogen atom. It also showed that a hydrogen atom was lost from C-24 during the formation of ophiobolin A from ophiobolin C, suggesting that a diene was involved in the process of forming the ether.

5.138          5.139

## 5.7 Fungal Triterpenoids and Steroids

Squalene (**5.141**) is the parent hydrocarbon of the triterpenoids and steroids. It is formed by joining two $C_{15}$ farnesyl diphosphate units (**5.140**) together in a head-to-head manner as shown in Scheme 5.13. In this scheme one mevalonoid hydrogen atom at C-11/12 is replaced by a hydrogen atom from NADPH. The triterpenes are then formed by the enzymatic cyclization of squalene, which may be initiated by either a proton or, as in most cases, by oxidative attack and the intervention of squalene epoxide (**5.142**). Most fungal, as opposed to bacterial, triterpenes belong to the tetracyclic class and are formed (Scheme 5.14) by the cyclization of squalene epoxide (**5.142 → 5.143**). Although there is an apparent similarity between the cyclization of squalene epoxide and the initial cyclization of geranylgeranyl diphosphate leading to the cyclic diterpenes, the stereochemistry of the reaction is different, particularly in the formation of the C-9–C-10 bond. Much of the information on triterpenoid and steroid biosynthesis was initially derived from mammalian systems and then extended to studies with microorganisms, particularly yeasts. Studies on the cyclase from the yeast *Saccharomyces cereviseae* have been particularly informative on the role of amino acid residues in stabilizing intermediates in the cyclization.

There are some significant differences between fungal triterpenes and sterols and those found in the plant, bacterial and mammalian kingdoms. Unlike higher plants but in common with mammals, lanosterol rather than cycloartenol is the parent of the tetracyclic triterpenes. Pentacyclic triterpenes are very rare in fungi. Alkylation of the side-chain at C-24 with the introduction of an extra methyl group is a very common process with fungi, leading, for example, to the modified side-chain of ergosterol. Ergosterol, unlike cholesterol, retains

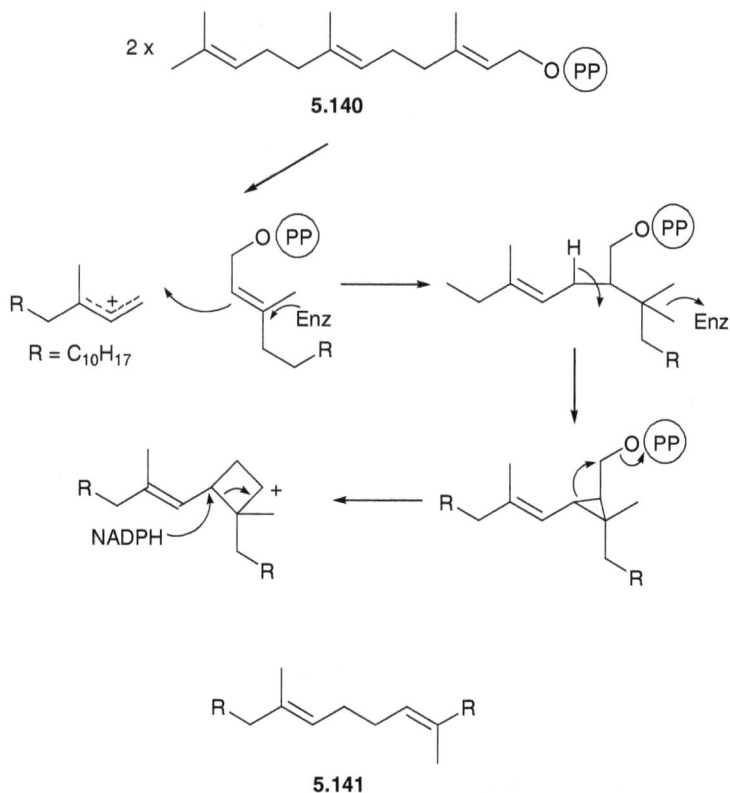

**Scheme 5.13**   Biosynthesis of squalene.

the ring B diene and this leads to a range of oxidation products. Whereas plants produce a biologically-active range of spiroketals and their glycosides known as the sapogenins and saponins, these have not been found as fungal metabolites. The C-20–C-22 oxidative cleavage of the fungal side-chain to generate $C_6$ and $C_7$ fragments and pregnane is reminiscent of mammalian metabolism rather than bacterial metabolism. Finally, fungi produce a family of metabolites with a carbon skeleton corresponding to the proto-lanostanes in which the backbone rearrangement of a methyl group and hydrogen atoms has not taken place.

## 5.7.1   Ergosterol

Just as cholesterol is a widespread mammalian sterol, which plays an important role in membrane structures, so ergosterol (**5.144**) and its esters are found in fungal cell walls. Lanosterol (**5.143**) has been established as its precursor (Scheme 5.15). The stage at which the alkylation at C-24 occurs depends on the organism. The extra carbon originates from methionine and, during alkylation, a rearrangement of a hydrogen atom from the C-24 of lanosterol to C-25 of ergosterol occurs. In some organisms 24-methylenelanost-8-en-3β-ol has been

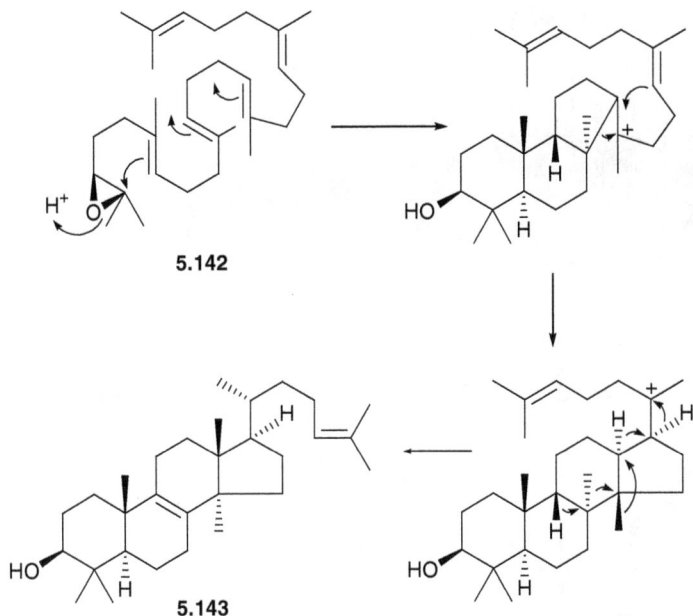

**Scheme 5.14** Cyclization of squalene epoxide.

identified whilst in others methylenation takes place at a later stage. The C-24(28)-alkene is reduced to generate the methyl group of ergosterol. The stereochemistry of the formation of the C-22 double bond in fungi involves the elimination of the (pro-*S*) hydrogens from C-22 and C-23.

The C-14 demethylation in ergosterol biosynthesis has been examined thoroughly as it is the target for the azole fungicides. The methyl group is oxidized to the level of a formyl group and then elimination occurs with the loss of the C-15α hydrogen to form a 14(15)-ene. Subsequent enzymatic reduction of this alkene takes place with the trans addition of hydrogen. Consequently, the hydrogen atom that once occupied the 15β position in lanosterol possesses the 15α stereochemistry in ergosterol. Isomerization of the $\Delta^8$ double bond to the $\Delta^7$-position and the formation of the ring B diene involve the loss of hydrogen from C-7β and from C-5α and C-6α.

The diene of ergosterol is very sensitive to oxidation. The 5α,8α-peroxide of ergosterol and a substantial number of other oxidation products have been found as fungal metabolites.

## 5.7.2 Fusidane Steroidal Antibiotics

Helvolic acid (**5.145**) was the first of the steroidal fusidane antibiotics to be isolated. It was obtained from *Aspergillus fumigatus* var$^n$ *helvola* in 1943 by Chain and Florey. Cephalosporin P$_1$ (**5.146**) was described by Abraham as a metabolite of the same Brotzu strain of *Cephalosporium acremonium* as that

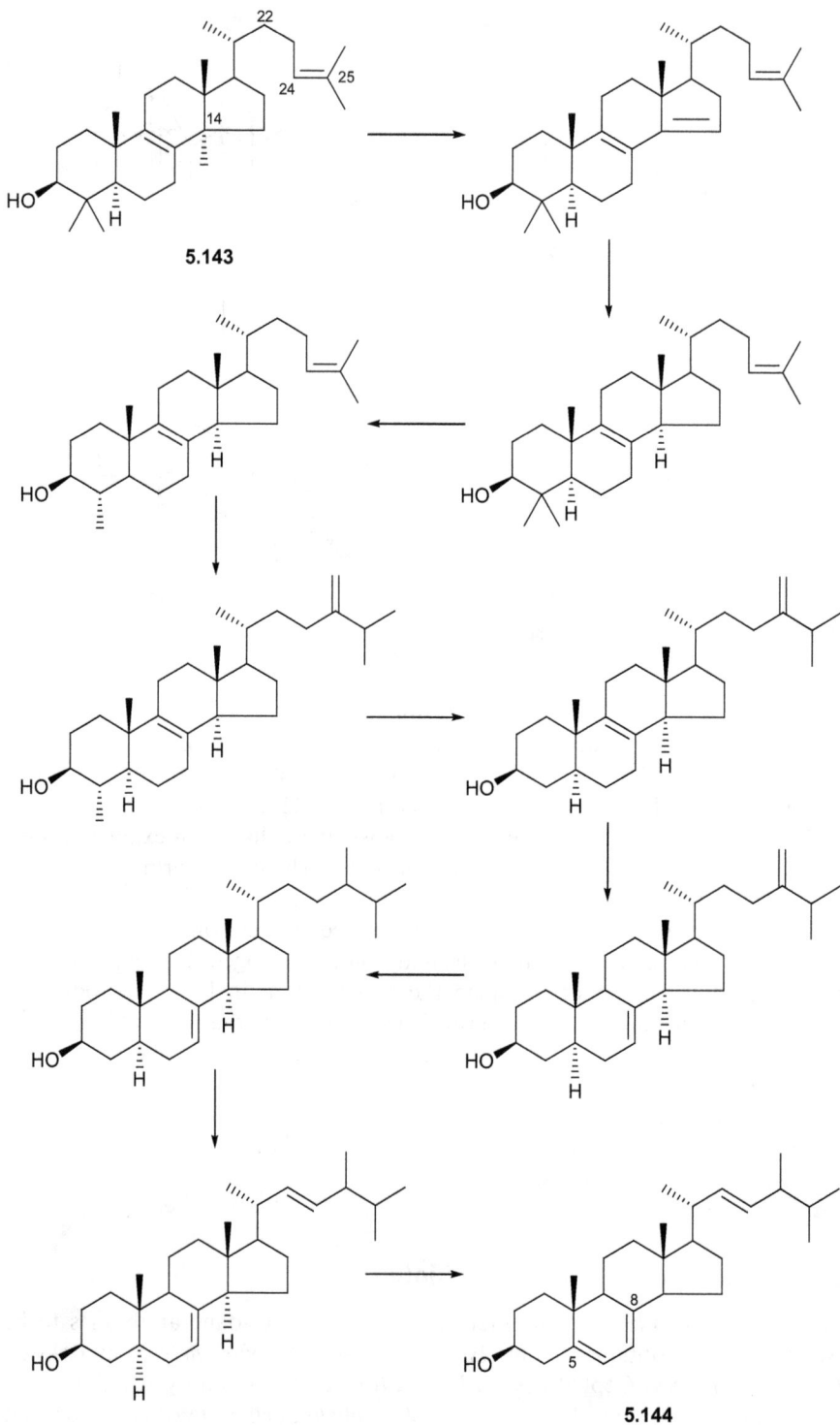

**Scheme 5.15** Biosynthesis of ergosterol.

which gave the cephalosporin β-lactam antibiotics. It was separated from the latter by its solubility in butyl acetate. Fusidic acid (**5.147**) was obtained as an antibiotic from *Fusidium coccineum* in 1962. Because of its commercial interest substantial quantities of fusidic acid were available and a structure was proposed in 1962 by Godtfredsen and Vangedal. This structure was modified and the stereochemistry was established in 1965. The structural work on these antibiotics was carried out at a time when steroid chemistry was being intensively studied in connection with the partial synthesis of steroid hormones. Hence there were many precedents available for the outcome of the reactions that were used in the degradation.

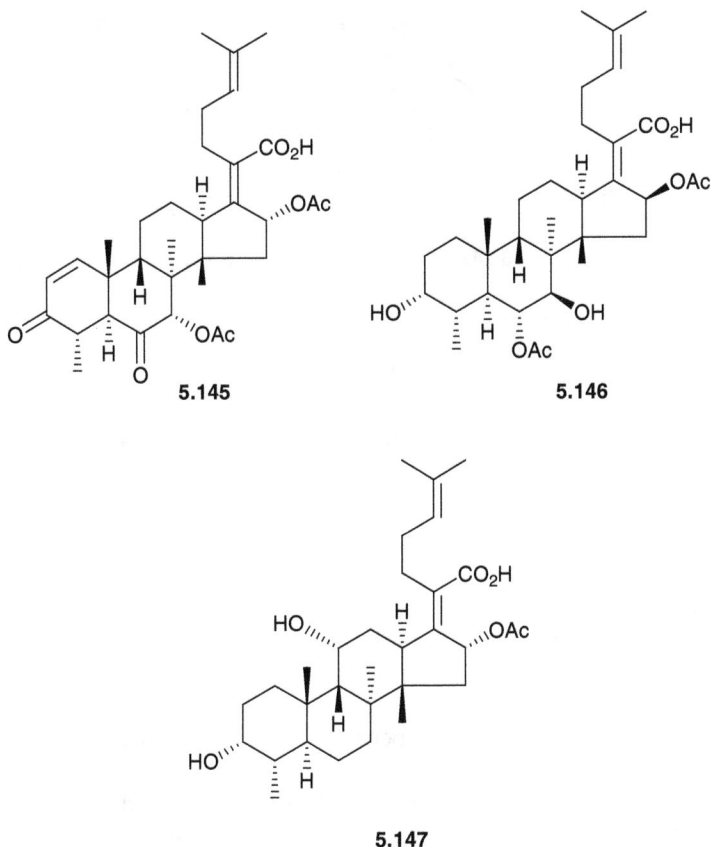

**5.145**                    **5.146**

**5.147**

Classical chemical degradation played a major role in the assignment of a structure to fusidic acid (**5.147**) whilst spectroscopic methods, mainly IR and $^1$H NMR, played a valuable supportive characterizing role. The underlying carbon skeleton was established by dehydrogenation to 1,2,5-trimethylnaphthalene, 1, 8-dimethylphenanthrene and to the naphthofluorene (**5.148**). The structure of the side-chain was established by ozonolysis of the methyl ester of dihydrofusidic acid. Reductive work up gave methyl 2-hydroxy-6-methylheptanoate (**5.149**). Acetone was obtained by ozonolysis of fusidic acid itself. The location of the functional groups and the relationship of the carboxyl group to the acetoxyl group on ring D were established by the changes in the IR spectra on hydrolysis, simple oxidation reactions and lactonization. The relationship of the ring C and ring D functional groups on the tetracyclic framework was established by oxidation and hydrolysis of the ozonolysis product to give an ene-1,4-dione (**5.150**). Although this was originally interpreted in terms of a ring C hydroxyl group at C-12, in the light of later work the position of this group was revised to C-11. The location of the methyl groups followed from the methylation pattern of the dehydrogenation products and from the $^1$H NMR spectra.

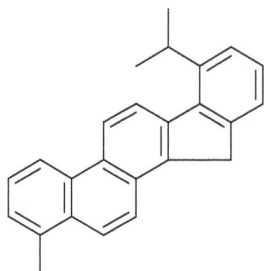

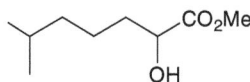

**5.148**                                                **5.149**

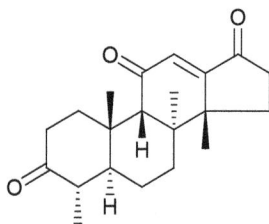

**5.150**

Evidence for the relative and absolute stereochemistry, which was reported in 1965, was based on a more detailed analysis of the $^1$H NMR spectra and on the signs of the circular dichroism curves of various ketones when compared to steroid models. The stereochemistry was finally confirmed by X-ray crystallography.

The original structural work on cephalosporin $P_1$ and helvolic acid in 1961 was incorrect because the presence of a tertiary methyl group was missed. It was only with the application of mass spectrometry and $^1$H NMR spectroscopy to the structural work that this omission was rectified. Ozonolytic removal of the side-chain similar to that used with fusidic acid and oxidation reactions of the tetracyclic core of cephalosporin $P_1$ established the presence of the α-glycol on ring B and its relationship to the ring A hydroxyl group. Again $^1$H NMR studies played an important role in defining the structure of cephalosporin $P_1$ (**5.146**) and the related helvolic acid (**5.145**).

Fusidic acid and helvolic acid were interrelated by an interesting microbiological strategy that has potentially wider applications. The fungus

*Acrocylindrium oryzae* was found to produce helvolic acid, *i.e.* it had the ability to introduce oxygen atoms onto the fusidane skeleton at C-6 and C-7 whereas *Fusidium coccineum*, which produces fusidic acid, does so at C-11. Incubation of 7-deacetoxy-l,2-dihydrohelvolic acid, obtained by reducing helvolic acid with zinc and acetic acid, with *F. coccineum* gave an 11α-hydroxy metabolite that on methylation and oxidation afforded a 3,6,11-triketone. Incubation of fusidic acid with *A. oryzae* gave a 6α-alcohol, which on methylation and oxidation led to the same triketone.

The biosynthesis of these antibiotics has attracted interest because of their protolanosterol structure. Thus, fusidic acid was shown to incorporate [2-$^{14}$C]- and (4*R*)-[2-$^{14}$C,4-$^{3}$H]mevalonic acid (**5.151**) as well as squalene in accordance with a pathway in which the final backbone rearrangement of lanosterol biosynthesis has not taken place.

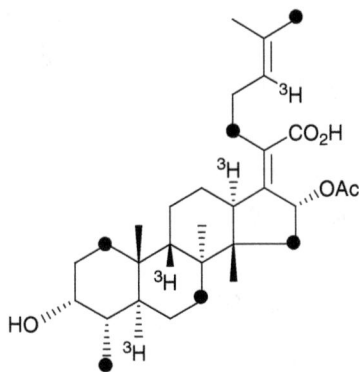

**5.151**

### 5.7.3 Viridin, Wortmannin and their Relatives

The steroidal antibiotics of the viridin series possess a selective antifungal action and they have an inhibitory effect on various steps in the cell-signalling process. Viridin (**5.152**) was first isolated in 1945 as an anti-fungal metabolite of *Glioclaudium virens* (*Trichoderma viride*). Preliminary studies showed that viridin was pentacyclic and that it contained a benzene ring, two easily reducible double bonds, three carbonyl groups, a hydroxyl group, a methoxyl group and a further inert ether oxygen atom. The structure of viridin followed from extensive oxidative degradation (Scheme 5.16), which gave benzene-1,2,3,4-tetracarboxylic acid (**5.154**), the phthalide **5.155** and the acid **5.156**. The latter was transformed into the naphthofuran **5.157**. The structures of these degradation products were established by synthesis. Careful examination of the $^{1}$H NMR spectrum of viridin finally led in 1966 to the structure. Demethoxyviridin (**5.153**) was isolated as an anti-fungal metabolite of *Nodulisporium hinnuleum* in 1975 whilst wortmannin (**5.159**), isolated in 1957, is an antifungal metabolite of *Penicillium wortmanii* and *Myrothecium roridum*. Their structures were established by a combination of spectroscopic studies and X-ray crystallography.

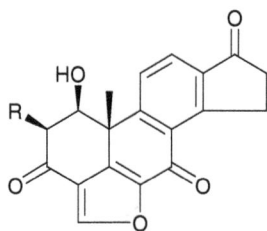

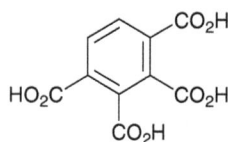

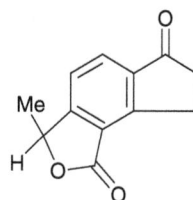

**5.152** R = OMe
**5.153** R = H

**5.154**

**5.155**

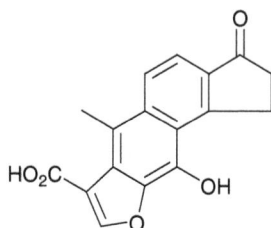

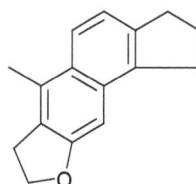

**5.156**

**5.157**

**Scheme 5.16**   Degradation of viridin.

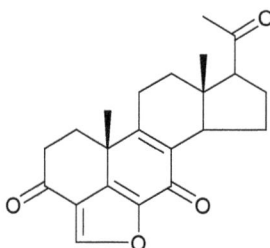

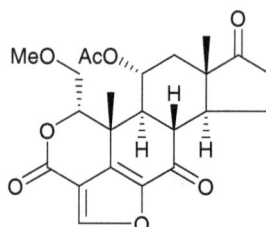

**5.158**

**5.159**

The furan ring of these metabolites, which is subject to the electron-with-drawing effect of the carbonyl groups at C-3 and C-7, is very sensitive to nucleophilic addition. The addition of biological nucleophiles to this centre has been associated with the biological activity of these compounds.

The labelling pattern of viridin biosynthesized from [2-$^{14}$C]mevalonic acid and of wortmannin biosynthesized from [1,2-$^{13}$C$_2$]acetate was consistent with a triterpenoid/steroidal origin. More detailed studies with demethoxyviridin biosynthesized from [1-$^{13}$C]-, [1,2-$^{13}$C$_2$]-acetate, [2-$^{13}$C]- and [5-$^{13}$C]mevalonate and on the location of the hydrogen atoms originating from acetate and the 2-, 4- and 5-positions of mevalonate defined the isoprene units and were again consistent with the triterpenoid origin of these metabolites. The 'extra' carbon atom of the furan ring at C-4 was shown to come from the C-4β methyl group of a proto-lanosterol precursor. The retention of three deuterium labels from acetate on the C-19 methyl group supported the intervention of a lanostane rather than a cycloartenol precursor which would have been a characteristic of plant sterol biosynthesis. The loss of a hydrogen atom from C-15 suggested that the removal of the C-14 methyl group followed a similar pattern to that found in other steroid pathways.

When a group of alcohols, *e.g.* **5.160**, which had been isolated from the fermentation broth of *Nodulisporium hinnuleum*, were biosynthesized from [2-$^{14}$C]mevalonic acid they had a similar specific activity to that of their co-metabolite, demethoxyviridin. These fragments represent the steroidal side-chain. However, the presence of the additional methyl group and its chirality suggests that the intermediate, which is cleaved, may have an ergostane rather than a cholestane side-chain. The isolation of virone (**5.158**) with a pregnane carbon skeleton is of interest in this metabolic step. The incorporation of squalene and lanosterol into these metabolites has been reported.

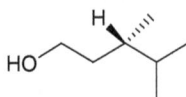

**5.160**

## 5.7.4   Triterpenoids of the Basidiomycetes

A group of closely related tetracyclic triterpenoids has been isolated from the higher fungi. In a series of papers published between 1904 and 1935, Zellner reported on the components of several higher fungi, including some *Polyporus* species. A common constituent was fungisterol (ergost-7-en-3β-ol) but several other crude products were obtained, including some that were probably tri-terpenoids. A systematic study by Jones of the birch-tree bracket fungus *Polyporus* (*Piptoporus*) *betulinus*, which commenced in 1940, led to the isolation of the polyporenic acids A (**5.161**), B and C. The study resumed after the war and the full structures were published in 1953–4. The structures of pinicolic acid A from *P. pinicola* found on pines and of eburicoic acid (**5.162**) from *P. anthracophilus* found on a *Eucalyptus* species were established at the same time. The work on eburicoic acid was carried out by Robertson in Liverpool. The elucidation of the structure of polyporenic acid A (**5.161**) benefited from experience gained in studies on the structure of lanosterol and on the penta-cyclic triterpenes. The identification of some structural features also showed the developing role of UV and IR spectroscopic methods in the 1950s. However, the purification of these compounds was impeded by the difficulty of separating the 8(9)-enes from the 7(8),9(11)-dienes that often accompanied them. Initial studies on polyporenic acid A revealed the presence of a carboxylic acid, two hydroxyl groups and two double bonds, each of differing reactivity. One double bond was a terminal methylene since formaldehyde was obtained on ozonolysis whilst the other was more hindered. Typical of the tetracyclic triterpenes, dehydrogenation with selenium gave 1,2,8-trimethylphenanthrene, affording an indication of part of the carbon skeleton. The location of the tetrasubstituted double bond followed precedents established with lanosterol. Oxidation with chromium trioxide gave, sequentially, the 7-keto-8-ene and then the 7,11-dione, and, if the 12-hydroxyl was not acetylated, the 7,11,12-trione. Commensurate with its position at C-12, the acetate underwent elimination to form a diene-7-one (**5.164**). The wavelength of the absorption maximum in the ultraviolet spectrum played an important part in the identification of the substitution pattern of these products.

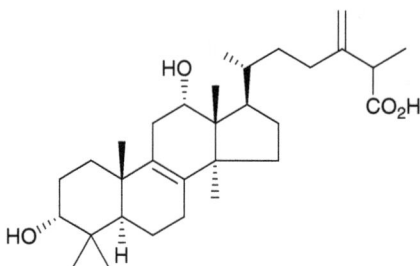

**5.161**

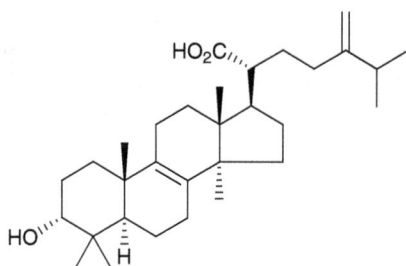

**5.162**

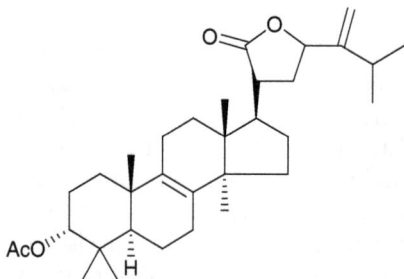

**5.163**

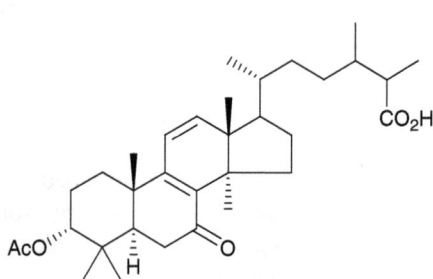

**5.164**

The position of the carboxyl group in the side-chain followed from its easy decarboxylation as a β,γ-unsaturated acid and from the isolation of acetaldehyde on cleavage of the alkene of the decarboxylation product. The hydroxyl group on ring A was shown to possess the less-common axial C-3α configuration.

Whilst this work was in progress comparable studies were carried out on eburicoic acid. In this case the relationship of the acid at C-21 to the alkene at C-24 was established by oxidation with SeO$_2$ to form a γ-lactone (**5.163**). The size of the lactone ring was established by its IR absorption, correlations that at that time had only recently been established. The degradation of eburicoic acid linked it to lanosterol *via* a common hydrocarbon, a standard strategy in natural product structure elucidation. Polyporenic acid C was interrelated with eburicoic acid.

Polyporenic acids A and C have anti-inflammatory activity and, in a bio-assay for potential anti-tumour activity, suppressed the oedema induced by 12-*O*-tetradecanoylphorbol-13-acetate.

An interesting feature of these triterpenes is the presence of the extra methylene at C-24 that is typical of ergosterol, indicating that this group may be introduced relatively early in the biosynthetic modification of lanosterol.

The fasciculols, *e.g.* fasciculic acid A (**5.165**), which were obtained from the bitter-tasting fruiting bodies of *Hypholoma* (*Neamatoloma*) *fasciculare* (sulphur tuft), are plant growth inhibitors. Some are esterified with a hydroxymethyl-glutarylglycyl unit. The compounds, particularly fasciculol C, are calmodulin inhibitors.

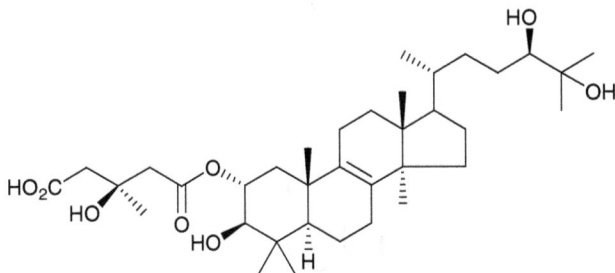

**5.165**

The fruiting bodies of *Ganoderma lucidum* are widely used in Oriental medicine for a range of conditions, particularly those involving inflammation and the immune system. In Chinese traditional medicine the fungus is known as Ling Zhi and in Japan as Reishi. Over 50 different tetracyclic triterpenes have been isolated. The role of 1D and 2D-$^1$H and $^{13}$C NMR techniques in estab-lishing their structures contrasts with the chemical degradative work used some 50 years earlier in determining the structures of the polyporenic acids. These structures are exemplified by that of ganoderic acid A (**5.166**). These oxygen-ated lanostanes have effects on many enzyme systems. Thus, 7-oxoganoderic acid Z inhibits HMG-CoA reductase whilst its decarboxylation products affect later stages in cholesterol biosynthesis. Ganoderic acid X is cytotoxic and in-hibits topoisomerase and can induce apoptosis in cancer cells. Ganoderic acid F possesses inhibitory activity against the angiotensin converting enzyme, pos-sibly accounting for the anti-ulcer activity of the fungus. Other members of the series have shown activity against HIV-1 protease, the DNA polymerases and in a primary screen for activity against tumour promoters using the induction of the Epstein–Barr virus early antigen. In this connection there are reports in Chinese medicine of the use of extracts of *G. lucidum* in conjunction with cancer chemotherapy. However, notably, the fungus also produces bio-active poly-saccharides. Not surprisingly there have been considerable efforts to grow this organism in stirred culture.

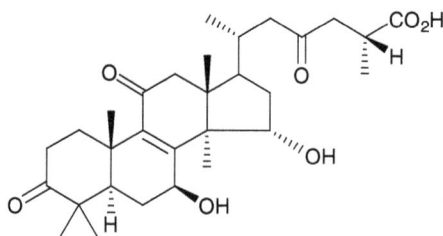

**5.166**

Another fungus, *Poria cocos*, which is also used in Chinese traditional medicine, contains a range of hydroxylated lanostanes, including the poricoic acids, *e.g.* poricoic acid A (**5.167**), which inhibit the tumour promoting effects of 12-*O*-tetradecanoylphorbol-13-acetate and possess cytotoxic effects against human cancer cell lines.

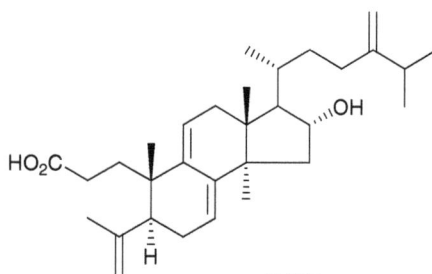

**5.167**

The carotenoids which are tetraterpenoids, are discussed in the chapter on fungal pigments (Chapter 7).

## 5.8 Meroterpenoids

Although meroterpenoids were originally defined as natural products containing terpenoid fragments along with structures of different biosynthetic origin, a more limited definition is that they are natural products of mixed terpenoid : polyketide origin. This has come to be accepted. Meroterpenoids include several fungal metabolites, many of which have been isolated from *Aspergillus*, *Penicillium* and *Helminthosporium* species. The commonest group are derived from triprenylphenols. The simpler members of this group are exemplified by grifolin (**5.168**) from *Grifolia confluens* and the siccanochromenes (**5.169**) and siccanin (**5.170**) from *Helminthosporium siccans*. However, in some cases both the terpenoid and the polyketide moiety have undergone substantial structural modification and their biosynthesis has required a considerable effort to unravel.

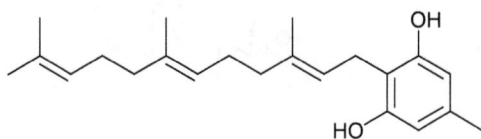

**5.168**

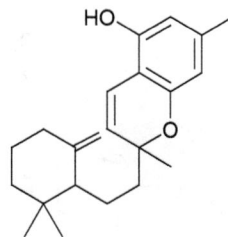

**5.169**

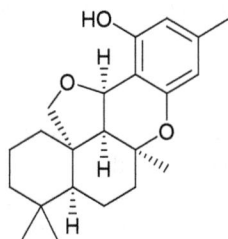

**5.170**

The combination of a lipophilic terpenoid fragment with a polar polyketide moiety has produced structures with significant biological activity. Thus the pyropenes A–R, *e.g.* **5.171**, from a strain of *Aspergillus fumigatus* are powerful inhibitors of acyl CoA-cholesterol acyl transferase, an enzyme that contributes to the absorption of dietary cholesterol and the accumulation of cholesterol esters. These are problems that lead to atherosclerosis.

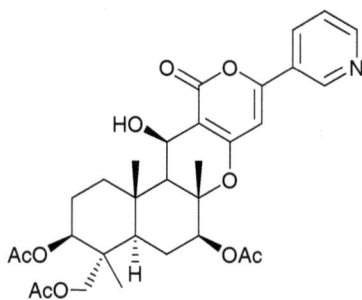

**5.171**

Andibenin A (**5.172**) and andilesin A (**5.173**), which were isolated from a strain of *Aspergillus variecolor*, are examples of these more complex structures. The structures were established by a combination of spectroscopic and X-ray crystallographic studies.

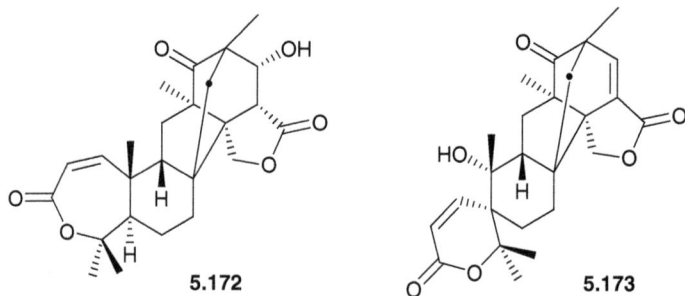

**5.172**          **5.173**

Since these compounds possessed a $C_{25}$ molecular formula and the structures had clear terpenoid characteristics, it was at one time suggested that they might constitute a group of sesterterpenoids. However, the enrichment and coupling pattern of andibenin B biosynthesized from [1-$^{13}$C]-, [2-$^{13}$C]- and [1,2-$^{13}$C$_2$] acetate made this proposal untenable. In particular, two methyl groups were not enriched by acetate units but were labelled by [methyl-$^{13}$C]methionine. This led to a biosynthetic scheme involving farnesyl diphosphate and 3,5-dimethyl-orsellinic acid. In confirmation of this 3,5-dimethylorsellinic acid was efficiently incorporated into andibenin A. Interestingly, orsellinic acid itself was a poor

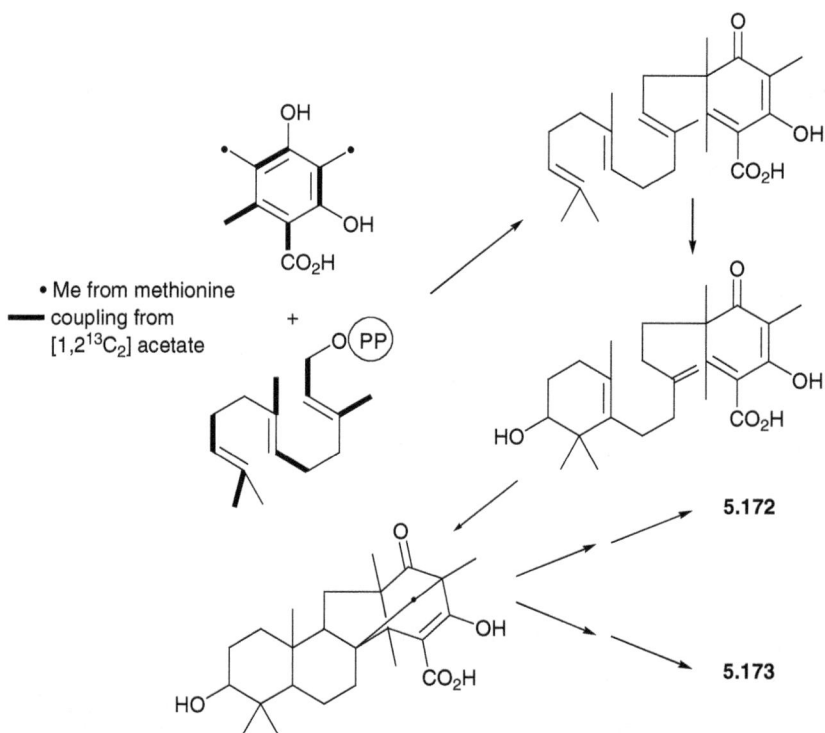

**Scheme 5.17** Biosynthesis of andibenin A (**5.172**) and andilesin A (**5.173**).

precursor, suggesting that methylation of the polyketide chain takes place before the aromatic ring is formed. On the other hand the terpenoid unit is attached after the aromatic ring is formed. The prenylation of the aromatic ring converts it into a dienone. An alkene may be formed when the farnesyl unit undergoes cyclization. A Diels–Alder reaction between this alkene and the diene may then generate the complex ring system that eventually leads, after further oxidation and rearrangements, to andibenin B. Experiments based on the incorporation of $[1\text{-}^{13}C,^{18}O_2]$acetate and $^{18}O_2$ revealed the origin of the oxygen atoms of andibenin B whilst the incorporation of deuterium from variously deuteriated mevalonates provided information on the mechanism of formation of the spirolactones. The pathway set out in Scheme 5.17 accommodates these results.

CHAPTER 6

# Fungal Metabolites Derived from the Citric Acid Cycle

## 6.1 Introduction

The citric acid (**6.1**) cycle (Scheme 6.1), also known as the tricarboxylic acid or Krebs cycle, is a major metabolic pathway of primary metabolism found in living systems. In this pathway acetyl co-enzyme A, derived by the catabolism of sugars and lipids, is oxidized to carbon dioxide and the energy that is released is stored in the formation of GTP, NADH and $FADH_2$. Some of the components in this pathway form the 'building blocks' for important amino acids such as aspartic acid and glutamic acid. In fungi a group of secondary metabolites are also derived from these acids, particularly oxaloacetic acid (**6.2**). The condensation of acetyl co-enzyme A with oxaloacetic acid and the decarboxylation steps illustrate two of the general metabolic reactions of citric acid intermediates that play an important role in their utilization to form secondary metabolites.

## 6.2 Citric Acid and Related Acids

The condensation between oxaloacetic acid (**6.2**) and acetyl co-enzyme A to form citric acid is a key step in the pathway. Citric acid (**6.1**) was originally isolated from citrus fruits and during the nineteenth century it was obtained from this source. However, the production of citric acid by a *Penicillium* species was observed by Wehmer in 1893. In 1916 Thom and Currie reported that strains of *Aspergillus niger* produced citric acid under conditions of mineral deficiency. During the 1920s the commercial production of citric acid was switched from citrus fruits to fermentation. Thus a method for citric acid production was reported using crude carbohydrate materials and the addition of

The Chemistry of Fungi
By James R. Hanson
© James R. Hanson, 2008

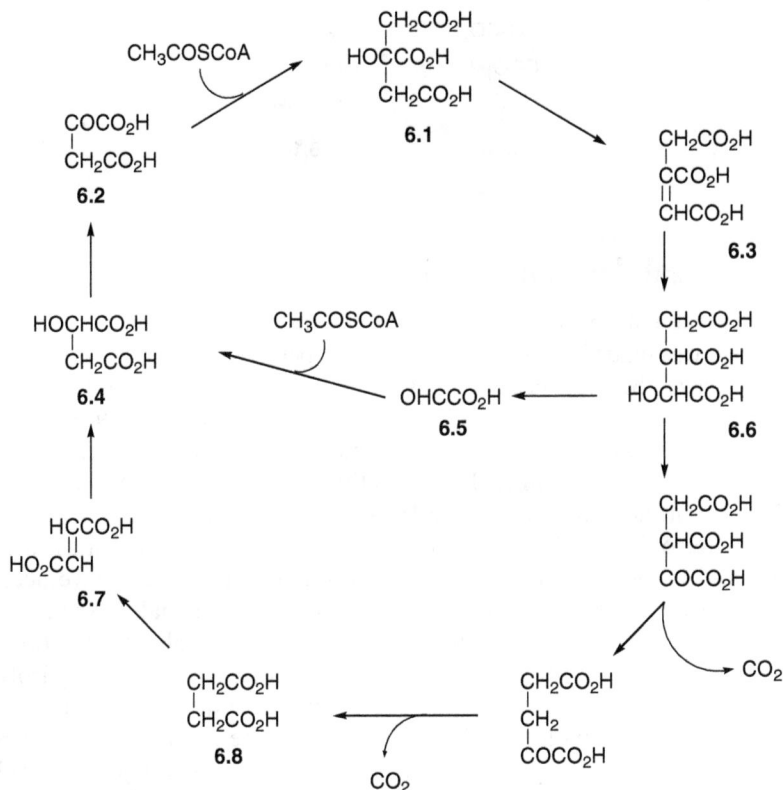

**Scheme 6.1**   The citric acid cycle.

potassium ferrocyanide as an inhibitors of parts of the citric acid cycle. The citric acid was recovered as its insoluble calcium salt. A modification of the citric acid cycle allows it to function not only as a cycle but to provide excess citric acid. Fission of one molecule of isocitric acid (**6.6**) may generate a molecule of succinic acid (**6.8**) and a molecule of glyoxylic acid (**6.5**). Whilst the succinic acid continues round the citric acid cycle, the glyoxylic acid may react with acetyl co-enzyme A to form malic acid (**6.4**) and thence oxaloacetic acid (**6.2**). Thus the citric acid that is lost from the cycle may be replenished by the 'glyoxalate' shunt.

Fumaric acid (**6.7**) is another component of the citric acid cycle that has been produced commercially by microbiological methods. *Rhizopus nigricans* has been used for this purpose and again the composition of the trace metal salts in the medium were important. Epoxysuccinic acid has been detected as a metabolite of *Aspergillus fumigatus*. Itaconic acid (**6.9**), which is formed by the decarboxylation of the citric acid cycle component, aconitic acid (**6.3**), is a further acid that has been produced commercially by fermentation using strains of *Aspergillus terreus*. A hydration product, itatartaric acid (**6.10**), has also been obtained from this organism.

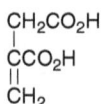

**6.9**

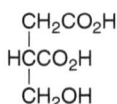

**6.10**

## 6.3 Fungal Tetronic Acids

Before its true nature as a vitamin deficiency disease was firmly established, a connection was made between the incidence of pellagra and the consumption of mouldy maize. Several *Penicillium* species were isolated from this source. Examination of one of these, *P. charlesii*, by Raistrick between 1933 and 1935 led to the isolation of a series of acidic metabolites that were derivatives of tetronic acids and exemplified by L-γ-methyltetronic acid (**6.11**), carolic acid (**6.12**), carolinic acid, carlosic acid (**6.14**) and carlic acid (**6.15**). The structure of L-γ-methyltetronic acid followed from its hydrolysis with dilute sulfuric acid to acetoin and carbon dioxide. Similar hydrolysis of carolic acid gave acetoin, carbon dioxide and γ-butyrolactone. Carolic acid was originally formulated as **6.13** but in the light of later NMR and X-ray crystallographic studies this was revised to **6.12**. In slightly acidic solution, the enol ether is easily hydrolysed and there is an equilibrium between the *E* and *Z* isomers of the enol ether. Dehydrocarolic acid, with a methylene in place of the methyl group of carolic acid, is a metabolite of *Penicillium cinerascens*. It was related to carolic acid by hydrogenation whilst the presence of the methylene was established by ozonolysis to give formaldehyde. The dibasic acids carlosic acid (**6.14**) and carlic acid (**6.15**) gave, on hydrolysis, two moles rather than one of carbon dioxide, acetoin and γ-butyrolactone or butyric acid, respectively.

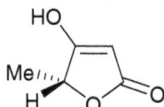

**6.11**

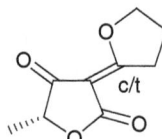

**6.12**

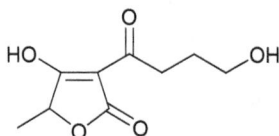

**6.13**

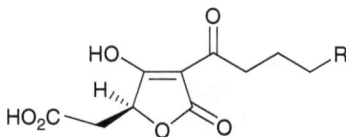

**6.14** R = H
**6.15** R = OH

Biosynthetic studies have shown that the butyric acid portion and the carbon atoms C-1 and C-2 of the tetronic acid are derived from acetate units whilst the

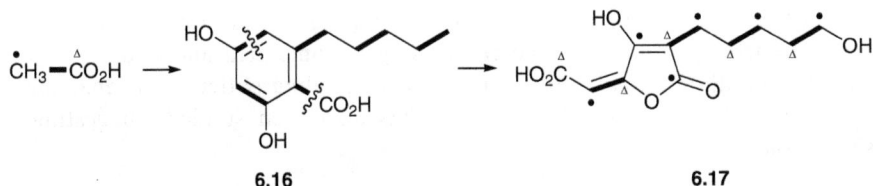

**Scheme 6.2**  Biosynthesis of multicolic acid.

carboxymethyl group and the carbon atoms C-3 and C-4 of the tetronic acid are derived from succinic acid, possibly *via* condensation of an activated methylene with malate or oxaloacetate.

Not all fungal tetronic acids are derived by this pathway. The biosynthesis of multicolic acid (**6.17**), a metabolite of *P. multicolor*, has been shown by carbon-13 labelling experiments to follow entirely a polyketide pathway. The proposed biosynthetic pathway (Scheme 6.2) involves the intermediacy of a 6-pentylresorcylic acid (**6.16**) and the cleavage of an aromatic ring.

A group of metabolites have been isolated, particularly from lichens, that may arise by condensation between the activated α-methylene of a fatty acid and the carbonyl group of oxaloacetate. Another example is agaricic acid from *Polyporus officinalis*, a metabolite with antibiotic properties associated with the agaricum of the Romans (see Chapter 1).

# 6.4   Canadensolide and Avenaciolide

Canadensolide (**6.18**) is an antifungal metabolite of *Penicillium canadense* that was reported in 1968. Its dilactone structure was established by a combination of spectroscopic and chemical methods. Thus the presence of the exocyclic methylene was suggested by the IR data and established by ozonolysis to give formaldehyde whilst the presence of the vicinal diol as part of the dilactone was indicated by the coupling pattern in the $^1$H NMR spectrum and confirmed by hydrolysis and cleavage of the resultant diol with sodium periodate to give valeraldehyde (pentan-1-al). There is a similarity between these metabolites and avenaciolide (**6.19**), a metabolite of *Aspergillus avenaceus*.

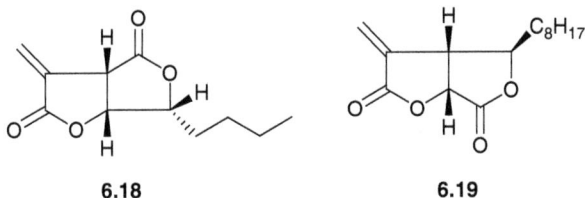

Biosynthetic studies with $[1,2^{-13}C_2]$acetate and $[2,3^{-13}C_2]$succinate have suggested that canadensolide may arise by the condensation of the α-methylene of a fatty acid and the carbonyl group of oxaloacetic acid to give a hexylcitric acid. Decarboxylation to the corresponding itaconic acid and hydroxylation would then afford canadensolide (**6.18**). The labelling pattern of avenaciolide (**6.19**) from [1-$^{13}$C]- and [2-$^{13}$C]acetate has also been studied, suggesting a similar biosynthesis.

## 6.5   Nonadrides

Glauconic acid (**6.20**) and byssochlamic acid (**6.21**) are unusual cyclic anhydrides that were isolated from *Penicillium glaucum* and *Byssochlamys fulva*. Their structures were established by chemical, spectroscopic and X-ray crystallographic studies. These structures suggest that they might be derived by the dimerization of two C$_9$ units, hence their name, nonadrides. Labelling studies have implicated the unsaturated anhydride (**6.24**) derived, *via* **6.23**, from an alkylcitric acid (**6.22**) in the dimerization. The labelled anhydride was incorporated in very high yield. The alkylcitric acid is formed by condensation of oxaloacetic acid and hexanoic acid. [2,3-$^{13}$C$_2$]Succinic acid has been specifically incorporated into the unit derived from oxaloacetic acid. The dimerization reaction (Scheme 6.3) is a base-catalyzed reaction to form glaucanic acid (**6.25**). Hydroxylation then affords glauconic acid.

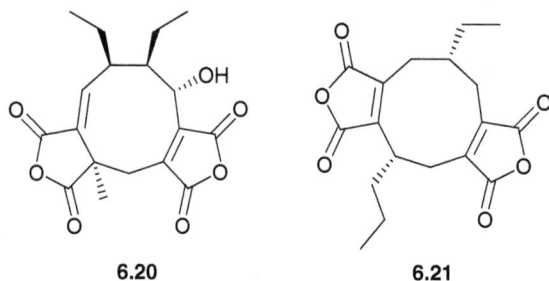

6.20                                    6.21

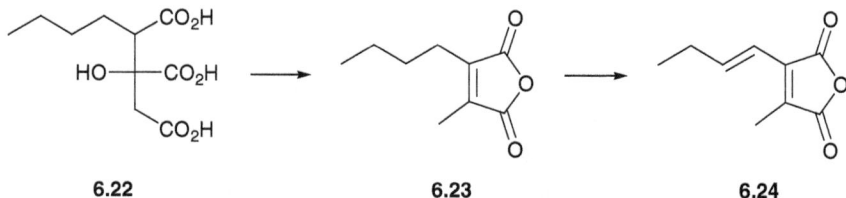

6.22                          6.23                          6.24

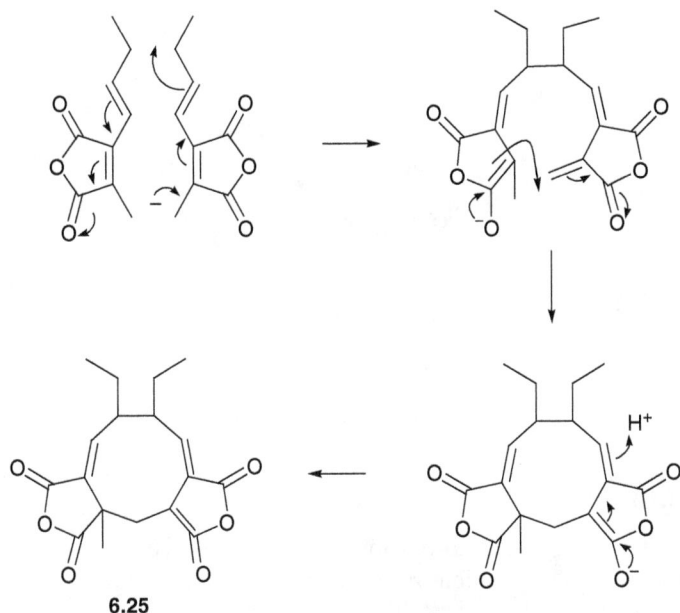

**6.25**

**Scheme 6.3**   Biosynthesis of nonadrides.

The structure of another nonadride, heveadride (**6.26**), obtained from the mycelium of *Helminthosporium heveae* was established from its spectroscopic properties and its oxidative degradation by potassium permanganate to a series of acids, the methyl esters of which were separated and identified by gas chromatography–mass spectrometry.

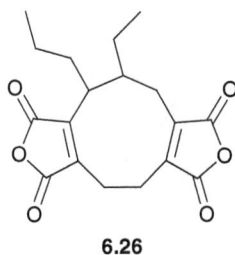

**6.26**

Strains of the fungus *Penicillium rubrum* growing on maize produce the rubratoxins A (**6.27**) and B, which were toxic to animals eating this maize. These compounds were subsequently detected as metabolites of *P. purpurogenum*. Their structures represent more highly oxidized forms of the nonadrides.

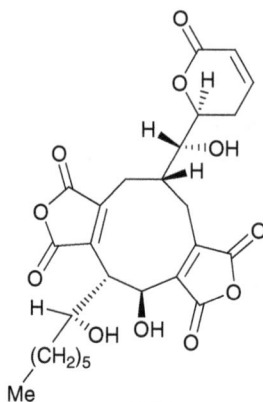

**6.27**

## 6.6   Squalestatins

The squalestatins, *e.g.* **6.28**, also known as the zaragozic acids, have attracted considerable interest as inhibitors of squalene synthase and hence of cholesterol biosynthesis and lipid deposition in the circulatory system. They are also inhibitors of farnesyl protein transferase and thus they may have other potentially useful biological applications. They are formed by *Phoma* species and also by *Setosphaeria khartoumensis*. The squalestatins are characterized by a dioxabicyclo-octane core bearing three carboxyl groups and two polyketide chains, one of which is attached as an ester. The biosynthetic incorporation of succinic acid into part of the bicyclo-octane, together with its oxygenation pattern, indicate that it may be derived *via* oxaloacetic acid. Both the polyketide chains have several pendant methyl groups attached to them, which arise from methionine, whilst benzoic acid acts as a starter unit for one of the chains. These complex structures are thus the summation of several biosynthetic pathways.

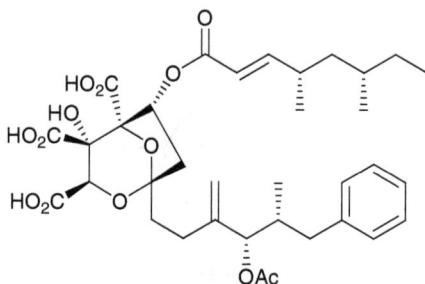

**6.28**

CHAPTER 7

# Pigments and Odours of Fungi

## 7.1 Introduction

The colours of fungi are one of the major characterizing features used in their identification. The red of the fly agaric, *Amanita muscaria*, the green of some *Penicillium* species and the deep violet of *Cortinarius violaceus* are distinctive features that exemplify the range of colours that have been observed. Furthermore, some fungi undergo distinctive colour changes when they are bruised or treated with alkali. The pigments of fungi, particularly of the fruiting bodies of the higher fungi, may play a role as attractants to insect vectors to facilitate the dissemination of spores. They are also light-absorbing molecules that can protect the organism from UV damage. A number have anti-bacterial activity and may protect the fungus from bacterial attack.

The structures of fungal pigments reveal major differences from those that are found in higher plants. Fungi do not possess chlorophyll or the anthocyanin pigments that dominate flower colours. Many of the pigments of fungi are quinones or similar conjugated structures. Some of the quinones are stored in the fungus as the less highly coloured quinol or phenol. These undergo aerial oxidation to the quinone when the fungus is damaged. In other cases the quinhydrone or a metal complex may be present.

The pigmentation of a culture may vary with its age. For example, *Penicillium chrysogenum*, the organism that produces penicillin, forms colonies on an agar plate which are initially white but which turn to a blue-green as the culture ages. The underside of the plate will appear quite yellow. An old culture will often be grey with white flecks. *Aspergillus niger*, a very common member of the Aspergilli, initially produces a white mycelium before turning yellow and forming black spores. Mature colonies then appear grey or greenish-black. The black pigment of many spores, 1,8-dihydroxynaphthalene : melanin, is formed by the laccase catalyzed oxidation of 1,8-dihydroxynaphthalene (**7.1**). This in turn is formed from the pentaketide, 1,3,6,8-tetrahydroxynaphthalene (**7.2**). The colour of a fungus can reflect its age, the extent of environmental stress or the exhaustion of a

The Chemistry of Fungi
By James R. Hanson
© James R. Hanson, 2008

component of the medium. In this chapter we consider the structures of the pigments of fungi within their biosynthetic classes.

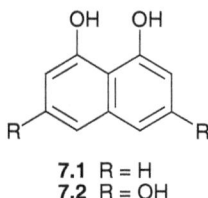

**7.1** R = H
**7.2** R = OH

# 7.2 Polyketide Fungal Pigments

Numerous quinones that are biosynthesized by the polyketide pathway have been isolated from fungi. The colour of the quinone that is isolated does not always reflect the colour of the fungus. This may be because the quinone is accompanied by its reduction products, forming a quinhydrone complex.

Much of the early work on these substances was carried out before the advent of modern physical methods of structure determination. Typical experiments involved reduction and deoxygenation with zinc dust to form identifiable aromatic hydrocarbons, *e.g.* 2-methylanthracene from a C-2 substituted anthraquinone. The oxygen functions were then placed on this carbon skeleton by unambiguous synthesis.

## 7.2.1 Fumigatin

A simple example is the quinone fumigatin (**7.3**), which was isolated from a culture of *Aspergillus fumigatus* in 1938. This particular strain first attracted attention when it was observed that the solution on which the fungus had been grown changed in colour from a yellowish-brown to a strong purple when it was made alkaline. Two grams of fumigatin were isolated from 35 litres of a culture medium of *Aspergillus fumigatus* that had been grown for a month. It was shown to contain one methoxyl group by a Zeisel determination (cleavage with hydrogen iodide) and one C-methyl group by a Kuhn–Roth determination (oxidation with chromium trioxide to acetic acid). The presence of one hydroxyl group was established since fumigatin formed a monoacetate and a monomethyl ether. The methyl ether was identical with synthetic 3,4-dimethoxy-2,5-toluquinone that was prepared for comparison purposes along with the other isomeric dimethoxytoluquinones. The location of the free hydroxyl group in fumigatin was established by converting it into the ethyl ether and showing that the corresponding quinol was different from synthetic 3-methoxy-4-ethoxytoluquinol. This had to be prepared by an unambiguous six-step synthesis from vanillin (**7.4**) of known structure. The acidity of the hydroxyl group accounts for the original observation of the change in colour on treatment with alkali. It is interesting to reflect that by using modern physical methods, including NMR experiments, this structure could probably be established in an afternoon while, given good crystals, an X-ray crystal structure determination would take about an hour.

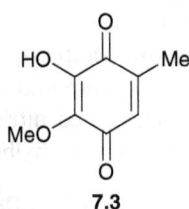

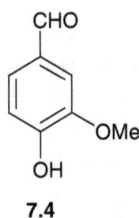

**7.3**          **7.4**

## 7.2.2  Auroglaucin and Flavoglaucin

In the 1930s and 1940s, Raistrick and a group of co-workers at the London School of Hygiene and Tropical Medicine carried out a systematic investigation of the pigments of various Aspergilli, Penicillia and *Helminthosporium* species. Species of the *Aspergillus glaucus* series (Latin *glaucus* = green), which are very common Aspergilli, are characterized by conidial heads that are a shade of green and hyphae with a colouration that varies through bright yellow to red. The red pigments sometimes crystallized on the mycelium. These fungi are found as spoilage organisms, *e.g.* on home-made jam and on some textiles. Extraction of the dried mycelium of 25 species in the *Aspergillus glaucus* series gave various amounts of three pigments: auroglaucin (**7.5**) as orange-red needles, flavoglaucin as lemon-yellow needles and rubroglaucin as ruby-red needles. In flavoglaucin the side-chain triene of auroglaucin has been reduced. Auroglaucin was bio-synthesized by a combination of polyketide and terpenoid pathways. Rubro-glaucin was eventually shown to be a mixture of the hydroxyanthraquinones physcion (**7.6**) and erythroglaucin (**7.7**).

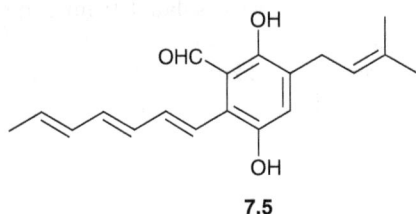

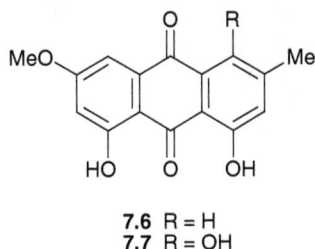

**7.5**

**7.6** R = H
**7.7** R = OH

## 7.2.3  Hydroxyanthraquinone Pigments

A further series of hydroxyanthraquinones were isolated by Raistrick at the same time from *Helminthosporium* species. These organisms are found on cereals and grasses. When *Helminthosporium gramineum*, the causative organisms of a 'leaf stripe' disease of barley, was grown on a Czapek–Dox medium it imparted a deep-red colour to the medium. Under some conditions the underside of the mycelial mat was partially covered with dark-red needle-shaped crystals. The main con-stituent was a trihydroxyanthraquinone, helminthosporin (**7.8**). Its structure was established by chemical means. The 2-methylanthracene was obtained by distillation over zinc dust in a stream of hydrogen and hence helminthosporin was a trihydroxy-2-methylanthraquinone. The triacetate was oxidized with chromic acid which converted the methyl group into a carboxylic acid. This underwent decarboxylation to the known 1,4,5-trihydroxyanthraquinone. The structural problem then hinged on relating the hydroxyl group and the methyl group. This was solved by vigorous oxidation with concentrated nitric acid, which gave, in addition to oxalic acid, 2,4,6-trinitro-5-hydroxy-3-methylbenzoic acid (**7.9**).

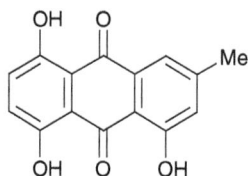 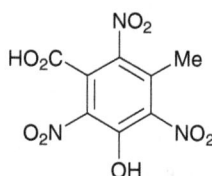

**7.8**                                      **7.9**

The identity of this degradation product was established by comparison with an authentic sample that was prepared by nitration of 5-hydroxy-*m*-toluic acid. The methyl group and the single hydroxyl group were therefore in the same ring. Helminthosporin was then assigned the structure of 2-methyl-4,5,8-trihydroxyanthraquinone (**7.8**). Despite the vigorous conditions, the absence of chromatographic methods of separation and the lack of spectroscopic information, quite complex structures were established in the 1930s and 1940s. It is again instructive to consider how this structure might be established by modern physical methods using NMR experiments.

Examination of 40 different *Helminthosporium* species established the widespread occurrence of this and related anthraquinone pigments such as catenarin (**7.11**) from *Helminthosporium catenarium*. An isomer of helminthosporin, islandicin (2-methyl-1,4,5-trihydroxyanthraquinone, **7.10**) was isolated from *Penicillium islandicum*. This particular organism was isolated from a mouldy specimen of skyr, which is a bacterially soured milk similar to yoghurt. Colonies of the fungus were characterized by a green conidial zone, near the margin, that was overgrown with orange to red hyphae in the central area. This pigment amounts to 3% of the dried mycelium.

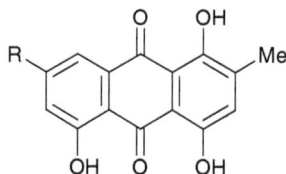

**7.10**  R = H
**7.11**  R = OH

## 7.2.4 Xanthone and Naphthopyrone Pigments

Not all the red pigments from the organisms were anthraquinones. A xanthone, rubrofusarin (**7.12**), has been obtained from several organisms. Many members of the genus *Fusarium* have a red pigmentation. The pigment aurofusarin (**7.13**), which is produced by *F. graminearum*, is a naphthopyrone dimer. It arises from a heptaketide that is coupled by a laccase.

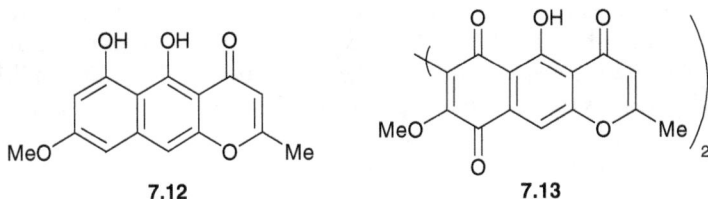

**7.12**          **7.13**

When the culture medium of *Fusarium oxysporum* is applied to colonies of other organisms such as *Aspergillus niger*, the hyphal tips cease to grow and morphological changes occur that are normally associated with senescence. A red pigment, bikhaverin (**7.14**), which was also obtained from *Gibberella fujikuroi* (*F. monoliforme*) when it was grown on a low-nitrogen medium, was responsible for this activity. The structure of bikhaverin was established by X-ray crystallography. The chemical properties of bikhaverin suggest that it may exist in solution as a mixture of two tautomers. Biosynthetically the compound is formed by cyclization of a nonaketide.

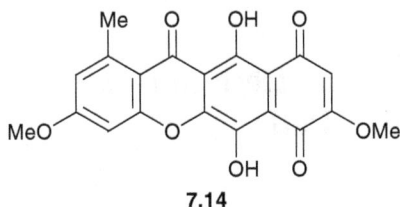

**7.14**

## 7.2.5 Extended and Dimeric Quinones

Extended quinones can contribute to the colours arising from fungal infections of plants. When the fungus *Chlorosplenium aeruginosum* (*Chlorociboria aeruginascens*) infects dead wood of deciduous trees such as the ash it imparts a characteristic green colour to the wood. The crystalline pigment xylindein (**7.15**) was first isolated in 1874 and structural studies were carried out in the 1920s. However, the structure was not finally established until 1962. The extended quinonoid structure arises by a dimerization.

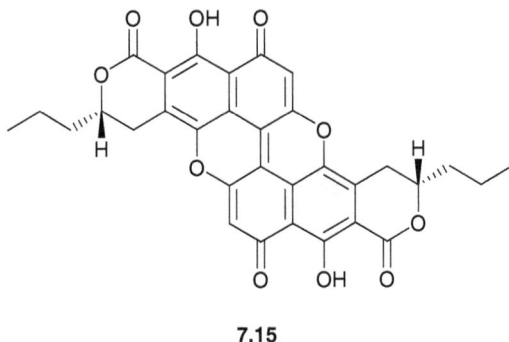

**7.15**

Higher fungi also produce pigments that are based on anthraquinone or anthrone skeleta. Several species of *Dermocybe* (*Cortinarius*) such as *Dermocybe cinnamomeolutea* and *Tricholoma, e.g. T. flavovirens,* produce a bright yellow intercellular pigment for which the dimeric structure flavomannin 6,6′-dimethyl ether, was established. Products of the oxidative coupling of anthrone are found in many *Cortinarius* species. Some of the structures are illustrated in **7.16**. The most widespread is the 7,7′-coupling as in **7.16b**. There are many variants based on different sites of coupling and O-methylation. There is also the possibility of hindered rotation about the Ar–Ar bond.

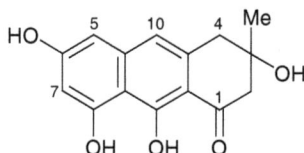

**7.16**  (a) 5,5′ = atrovirin
(b) 7,7′ = flavomannin
(c) 5,10′ = pseudophlegnacin
(d) 7,10′ = phlegmacin
(e) 10,10′ = tricolorin

# 7.3 Fungal Pigments Derived from the Shikimate Pathway

## 7.3.1 Terphenyls

Arylpyruvic acids derived from the shikimic acid $C_6$–$C_3$ biosynthetic pathway form the building block for many pigments of the higher fungi. The arylpyruvic acid units are joined together possibly by a carbanion process between the methylene ketone of one and the co-enzyme A ester of the other. The structure of grevillin A (**7.17**), isolated from *Suillus grevillei*, a bolete which grows on larch, illustrates this.

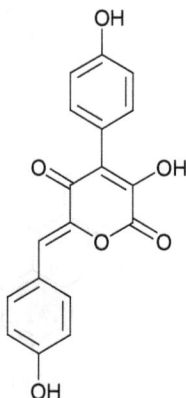

**7.17**

The quinone polyporic acid (**7.18**) is the parent terphenyl of several pigments. It is the major component of the bracket fungus, *Polyporus nidulans* (*Hapalopilus nidulans*), which grows on various deciduous trees, and can make up to 43% of the dry weight of the fungus. The brown outer skin of *Paxillus atromentosus*, a dark coloured fungus found on the damp stumps of conifers, contains a more highly oxidized compound, atromentin (**7.19**). The colourless reduced leuco form leucomentin is esterified with (2*Z*)-(4*S*,5*S*)-4,5-epoxyhexenoic acid. The same ester is present in the orange-yellow flavomentin pigments. Hydrolysis of these pigments leads to the conversion of the epoxyacid into an insect anti-feedant, osmundalactone (**7.20**). Since many of the quinones have anti-bacterial properties, the presence of an insect anti-feedant confers additional protective properties on these pigments.

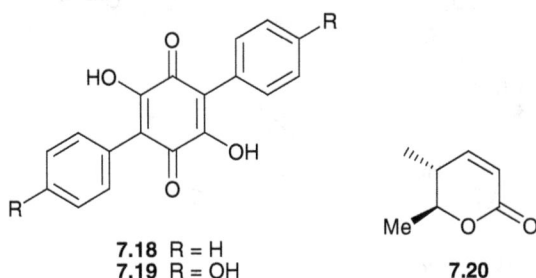

**7.18** R = H
**7.19** R = OH

**7.20**

Thelephoric acid (**7.21**) is a further oxidation product that is widespread amongst the Basidiomycetes. The delocalization of the phenolate anions forms the basis of the colour reactions that are observed when the fruiting bodies come into contact with alkali. The quinhydrones of corticins A (**7.22**) and B and their quinones give rise to the blue colour of *Corticium caeruleum*. The terphenylquinols are powerful anti-oxidants.

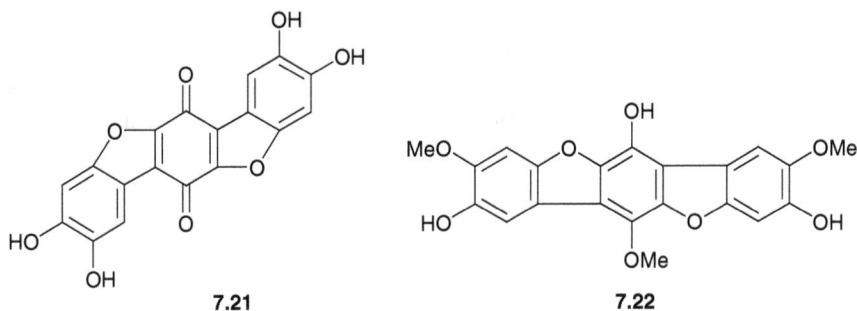

**7.21**

**7.22**

## 7.3.2 Pulvinic Acids

Cleavage of the hydroxyquinone ring system and relactonization affords another series of pigments known as the pulvinic acids, which may be exemplified by gomphidic acid (**7.23**). The dry-rot fungus, *Serpula lacrimans*, produces xerocomorubin (**7.24**), which is another example. These pulvinic acid moieties form part of a complex group of metabolites known as the badiones. They are produced by *Xerocomus badius*, an edible bay bolete found growing in spruce and pine woods. Norbadione (**7.25**) is found in quite high concentrations in another gasteromycete, *Pisolithus arrhizus* (*P. tinctorius*). These pigments attracted interest after the Chernobyl disaster. They bind potassium and caesium ions and were responsible for the quite high levels of the radionuclide $^{137}$Cs found in wild mushrooms after the explosion. A more complex member is a yellow pigment, sclerocitrin, which is found in the common spherical earth puff ball, *Scleroderma citrinum*.

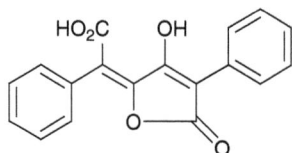

**7.23**

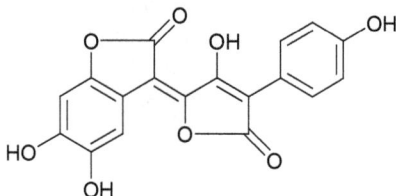

**7.24**

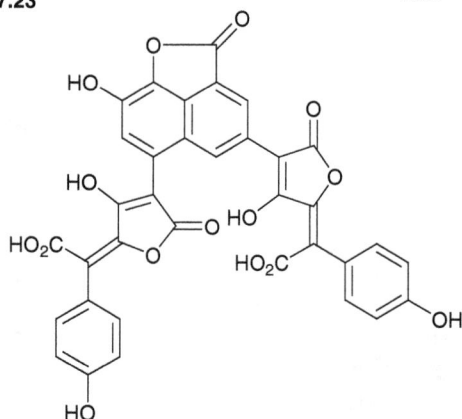

**7.25**

The orange-red caps of *Tricholoma aurantium* contain a pigment, aurantricholone (**7.26**), in which a pyrogallol ring attached to a pulvinic acid has been oxidatively dimerized to a purpurogallin analogue. This particular pigment occurs as a calcium complex.

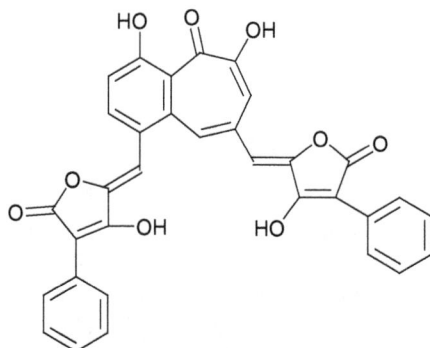

**7.26**

Ring contraction and decarboxylation of the central hydroxyquinone may produce a cyclopentenone ring, exemplified by involutin (**7.27**) and the pigment **7.28**. This unstable pigment is responsible for the brown stain that develops when the fruiting bodies of *Paxillus involutus* are bruised. This fungus is a common species that is found in beech and oak woods. Another *Boletus* is *Gyroporus cyanescens*, which is sometimes found on birch trees or in heathland and develops a blue colour when it is cut. This is associated with the related metabolite which is oxidized to the blue anion **7.30**. Protonation and hydration gives gyroporin (**7.29**).

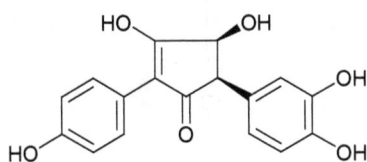

**7.27**

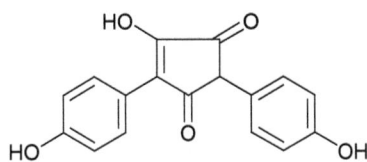

**7.28**

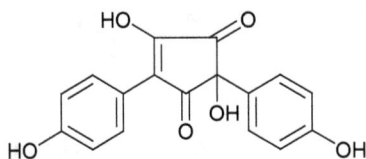

**7.29**

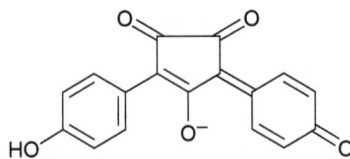

**7.30**

The 'tinder fungus' *Fomes fomentarius*, which grows on both beech and birch trees, gives a deep red colour in alkali. This is associated with the presence of fomentariol (**7.31**), a purpurogallin type of oxidation product of trihydroxy-cinnamyl alcohol.

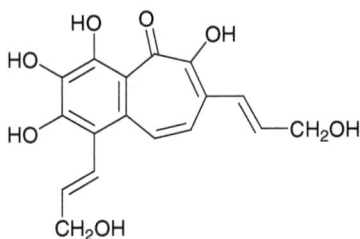

**7.31**

## 7.4 Some Pigments Containing Nitrogen

The striking red colour of the fly agaric *Amanita muscaria* arises from the presence of betalain pigments. These pigments, which are also responsible for the red colour of beetroots, are formed by the condensation of the amino group of various amino acids with the aldehyde of betalamic acid (**7.33**). This acid and

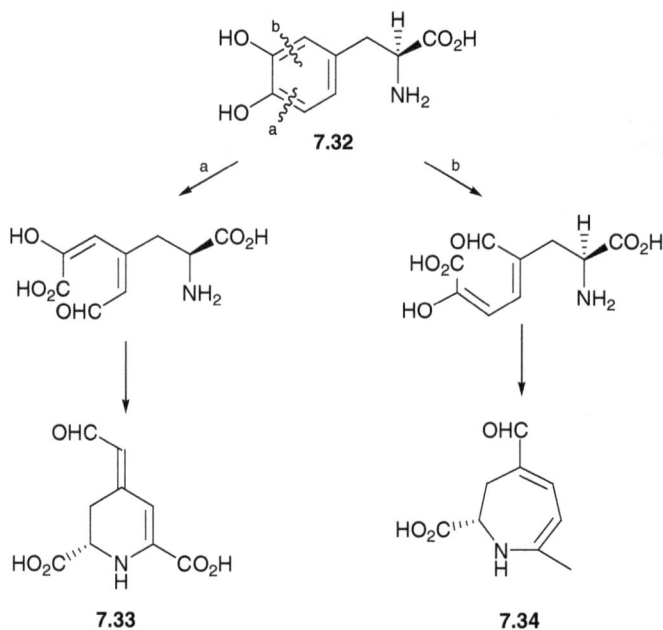

**Scheme 7.1**  Formation of secoDOPA pigments.

a yellow pigment, muscoflavin (**7.34**), are formed (Scheme 7.1) by cleavage of the $C_6$–$C_3$ aromatic amino acid L-3,4-dihydroxyphenylalanine (L-DOPA, **7.32**).

The fruiting bodies of several *Cortinarius* species have a remarkable deep violet colour. *Cortinarius violaceous*, which is found in deciduous woodland, can concentrate iron by as much as 100-fold over other typical Basidiomycetes. The violet colour has been ascribed to an iron-catechol complex containing dihydroxyphenyl-β-alanine (**7.35**). The latter is formed by the rearrangement of L-DOPA.

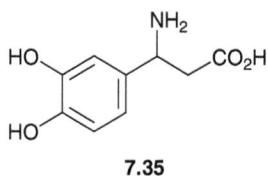

**7.35**

A characteristic of some Horse mushrooms (*Agaricus* species) is that they turn yellow when bruised. Examination of the 'yellow-staining mushroom' *Agaricus xanthoderma* has yielded the azaquinone agaricone (**7.36**), which is formed by aerial oxidation of a leucophenol in the damaged tissue.

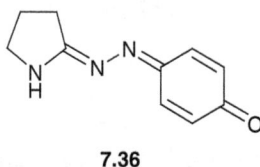

**7.36**

The wood-rotting fungus *Pycnoporus cinnabarinus* is found on dead branches of beech and birch. It has a cinnabar red appearance. The red pigment, cinnabarinic acid (**7.37**), is a phenoxazine-3-one that is biosynthesized by the oxidative dimerization of 3-hydroxyanthranilic acid. Like many of the pigments of the higher fungi, cinnabarinic acid has anti-bacterial properties.

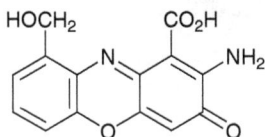

**7.37**

Green pigments are relatively rare amongst the higher fungi. One example is produced by a slimy milk cap, *Lactarius blennius*, which is found in European beech woods. The pigment blennione (**7.38**) of the grey-green fungus is also formed by coupling two dihydroxyanthranilic acid units together.

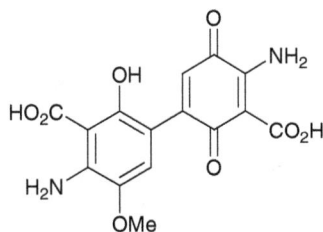

**7.38**

A combined $C_6$-$C_3$-polyketide pathway leads to the styrylpyrone hispidin (**7.39**). This pigment was originally isolated from *Inonotus hispidus*, a fungus found in clumps on fallen trees and amongst leaf litter. It and its relatives are widespread amongst *Gymnophilus*, *Hypholoma* and *Pholista* species such as *Hypholoma fasciculus* (sulphur tuft). The open-chain β-keto-ester **7.40** has been obtained as the pigment of a brick red Hypholoma *H. sublateritium*. These compounds are related to the substituted β-methoxyacrylates fungal metabolites known as the strobilurins (**7.41**). These anti-fungal compounds are produced by several Basidiomycetes and provide the lead compounds for a series of synthetic anti-fungal agents.

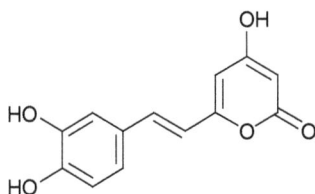

**7.39**

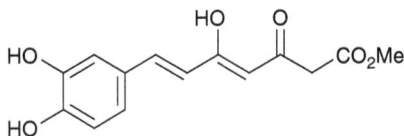

**7.40**

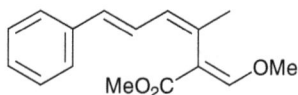

**7.41**

## 7.5   Terpenoid Pigments

Apart from the carotenoids relatively few terpenoids have sufficient conjugated unsaturation to contribute to fungal pigments. Exceptions are the sesquiterpenoid lactarazulenes, *e.g.* lactaroviolin (**7.42**), which are obtained from several *Lactarius* species of milk cap. These are discussed in Chapter 5.

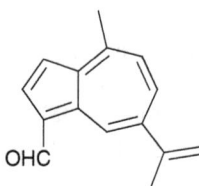

**7.42**

### 7.5.1   Fungal Carotenoids

Since fungi are non-photosynthetic organisms, carotenoids are not as widespread in fungi as they are in plants, where they play an important role in the photosynthetic process. Nevertheless, carotene hydrocarbons have been found in several fungi, including *Blakeslea trispora*, *Phycomyces blakesleanus* and *Neurospora crassa*. β- and γ-Carotene are the common hydrocarbons. The keto-carotenoid canthaxanthin (4,4'-diketo-3-carotene, **7.43**) has been isolated from *Cantharellus cinnabarinus* and *C. infundibilis*. Several *Cantharellus* species, such as the chantarelle (*C. cibarius*), have a yellow pigmentation that might arise from their carotenoid content.

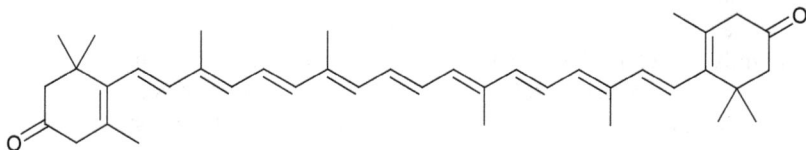

**7.43**

Whereas the isoprene units of the higher plant carotenoids that are biosynthesized within the chloroplasts are formed *via* the 1-deoxyxylulose pathway, the fungal carotenoids are biosynthesized from acetate *via* mevalonic acid. The $C_{40}$ carbon skeleton of the carotenoids is formed by the head-to-head coupling of two $C_{20}$ geranylgeranyl diphosphate units by phytoene synthase.

Although there is a similarity to the formation of the $C_{30}$ hydrocarbon squalene, there is a major difference in that the dimerization forms an alkene, phytoene (**7.44**). Stepwise dehydrogenations *via* phytofluoene (**7.45**) and neurosporene afford lycopene (**7.46**). The stereochemistry of the dehydrogenation of phytoene involves the loss of the pro-(2S)-and pro-(5R)-hydrogens of mevalonic acid. These dehydrogenation steps are inhibited by diphenylamine. Cyclization of lycopene affords β-carotene. Modification of γ-carotene by *Neurospora crassa* by removal of a terpene unit affords the pigment, neurosporaxanthin. Hydro-xylation and oxidation of β-carotene affords astaxanthin (3,3′-dihydroxy-β-carotene-4,4′-dione). Interest in this metabolite of a yeast, *Pfaffia rhodozyma*, has centred on its application as a pigment supplement for salmon.

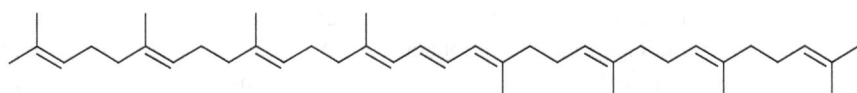

**7.44**

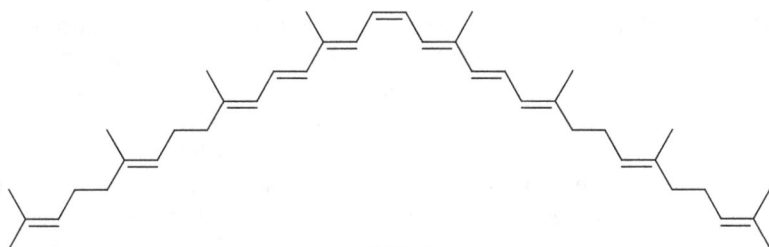

**7.45**

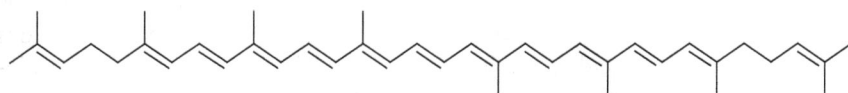

**7.46**

Heterothallic fungi of the order Mucorales display a simple form of sexuality mediated by inter-diffusing hormones. There are two forms of each organism, the (+) and (–) sexual forms. If these two strains of opposite sex are growing in close proximity, they mutually elicit the first morphogenetic step in the sexual process, the formation of zygophores. The induction of zygophore formation is mediated by trisporic acids, *e.g.* **7.47**, the most active of which is 9-*cis*-trisporic acid B. The trisporic acids are cleaved (apo) carotenoids and are formed *via* prohormones. The prohormones from the (+) cultures are converted into trisporic acids by the (–) cultures and conversely the prohormones from the (–) cultures are converted by the (+) cultures into trisporic acids. These prohormones occur in low concentrations (*ca.* $1 \, mg \, L^{-1}$). Nevertheless their structures were established and they were shown to be derived from β-carotene.

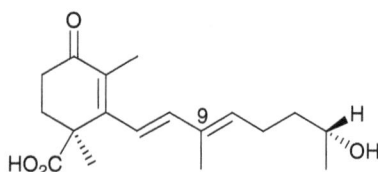

**7.47**

# 7.6   Lichen Substances

Lichens are symbiotic organisms that consist of a fungus and a photosynthetic alga or cyanobacterium. The alga or cyanobacterium provides the sugars for growth while the mineral nutrients are obtained from the water running over the surface to which the lichen is attached. Hence the growth of the lichen is very sensitive to environmental contamination. Lichen substances are often characteristic of the fungal (mycobiont) as opposed to the algal (phycobiont) component of the organisms. They can be produced in substantial amounts and their presence is an important contributory factor to the identification of the lichen. Various micro-scale mass spectroscopic and HPLC-UV techniques for the detection of lichen pigments have been developed, particularly in the study of lichens growing in an inhospitable environment. The pigments of some lichens have been associated with the accumulation of metal ions from the substrate to which the lichen is attached. Lichen substances have been extracted and modified to create pigments for cloth.

Lichens produce a series of dimeric tetraketides, based on orsellinic acid, known as the depsides. About 100 of these compounds have been isolated. They consist of an aromatic phenolic acid that is esterified by the phenol of a second, not necessarily identical phenolic acid. Several examples have an alkyl chain attached to one or both aromatic rings. They are exemplified by atranorin (**7.48**) from *Lecanora atra* and lecanoric acid (**7.49**), which occurs in numerous lichens, particularly from the Parmelliaceae. and Roccellaceae. Whereas lecanoric acid is a dimer of orsellinic acid, gyrophoric acid is a trimer of this acid. The biosynthesis of these compounds has been studied by suspending the lichen thallus in a nutrient medium. Related to these compounds are cyclic ethers known as the depsidones and a small group of phenol oxidation products known as the depsones. These are exemplified by diploicin (**7.50**) from *Diploicia canescens*, stictic acid (**7.51**) from the Parmelliaceae and picrolichenic acid (**7.52**) from *Pertusaria amara.*

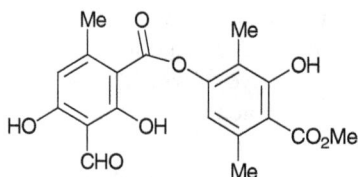

**7.48**

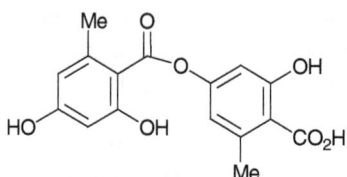

**7.49**

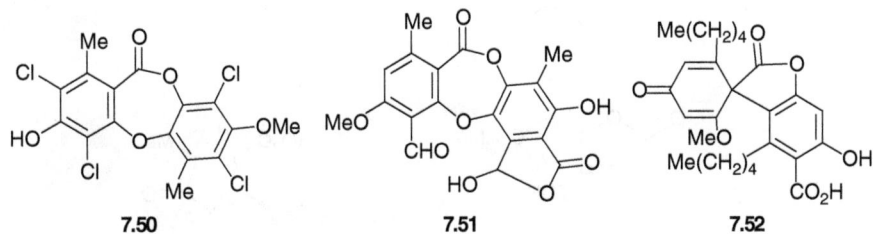

**7.50**          **7.51**          **7.52**

Usnic acid (**7.53**)—a highly biologically active dibenzofuran—is a phenol oxidation product of an acylphloroglucinol. It is produced by numerous lichens, including several *Usnea* species. Apart from its antibiotic and anti-leukemic activity, it has insect anti-feedant activity. Its presence may account for the observation that many lichens are resistant to serious insect attack.

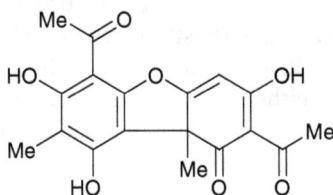

**7.53**

A second large group of polyketide metabolites that have been isolated from lichens are heptaketide xanthones. Their structures are mainly based on norlichexanthone (**7.54**). An interesting feature of these compounds is the presence of chlorine in several of the metabolites. Lichens also produce some anthraquinone pigments, exemplified by physcion (**7.6**) from *Xanthora* species.

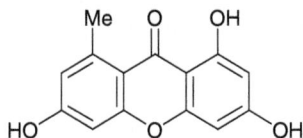

**7.54**

These aromatic compounds have strong ultraviolet absorption and may serve to protect the lichen from photochemical damage. They also have antibiotic properties and may protect the organism against bacterial decay. Their plant growth inhibitory properties may preserve the lichen in an ecological niche, preventing the growth of plants on a potential nutrient surface.

A further family of aromatic compounds produced by lichens is derived from shikimic acid *via* phenylalanine and is exemplified by vulpinic acid (**7.55**), which was isolated from *Letharia vulpina*. Vulpinic acid has been synthesized by the lead tetra-acetate oxidation of polyporic acid. Lichens also produce some alkylated citric acid derivatives such as lichesterinic acid (**7.56**) and some sterols.

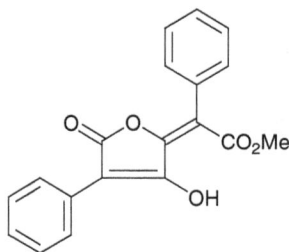
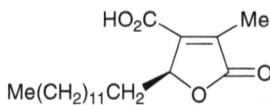

**7.55**                                    **7.56**

## 7.7 Odours of Fungi

When you enter a damp cellar, open a packet of mushrooms or go for a walk in the woods in the autumn, you will often smell the chemical activities of fungi. Although many of the more pungent smells of decay are associated with bacterial activity, fungi make a characteristic, often more 'musty', contribution. The volatile compounds of mushrooms contribute to their organoleptic properties. Other compounds behave either as insect attractants or as feeding deterrents. Yet others contribute to the relationship of a fungus with its competitors.

Whilst monoterpenes and simple aromatic compounds are produced by fungi, they do not have quite the dominance that is found in floral odours. Fungi produce different proportions of the common volatile components. Some organisms have a characteristic smell that can be used to identify them. The 'stink horn' *Phallus impudicus* and the 'aniseed toadstool' *Clitocybe odora* did not get their names for nothing. Dogs have been trained to detect the presence of the 'dry-rot' fungus *Serpula lacrimans* in buildings. Mass spectrometry coupled with 'electronic nose' technology has been developed to rapidly reveal the presence of unwanted fungal growth on foods such as cereals and bakery products. Variations in the volatile metabolites that were detected enabled predictions to be made concerning the identity of the infecting organism.

### 7.7.1 Organoleptic Components of Mushrooms

The volatile components produced by the cultivated mushroom, *Agaricus bisporus*, have been extensively studied. Almost 100 compounds have been detected by gas chromatography combined with mass spectrometry. The major compound, making up as much as 70% of the volatile material, is oct-l-en-3-ol (**7.57**). This compound has the characteristic mushroom smell and is produced by many fungi. The alcohol is chiral and one enantiomer, the (*S*)-form, is produced naturally, and although both enantiomers can be detected by man as having a mushroom-like odour, the threshold for detection of the (*S*)-enantiomer is lower. The corresponding ketone, oct-l-en-3-one, which also possesses a mushroom odour, has an even lower detection limit (0.004 ppm compared to 0.010 ppm for the alcohol). Our sense of smell is such that we are able to detect compounds at these very low concentrations. Indeed the cultivated mushroom does not produce very much volatile material—only a few milligrams per kilogram fresh weight. This poses problems of detection and analysis. Various entrainment techniques have been used to recover the volatile metabolites.

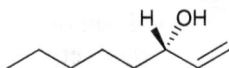

**7.57**

Other compounds that have been detected in mushroom volatiles include a range of $C_8$ alcohols, their esters and oxidation products and various other compounds such as benzyl alcohol, benzaldehyde, acetone, ethanol, ethyl acetate, 3-methylbutanol, 2,4-nonodienal and 2,4-decadienal together with, surprisingly, tetrachloro-1,4-dimethoxybenzene (drosophilin A methyl ether) (**7.58**).

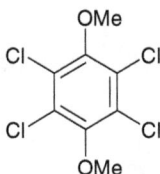

**7.58**

The major $C_8$ compounds are found in many fungal volatiles and are biosynthesized along with 10-oxo-*trans*-8-decenoic acid (**7.60**) and 13-hydroperoxy-*cis*-9-*trans*-11-octadecadienoic acid (**7.61**), by oxidation of linoleic acid (**7.59**). The fragmentation of the carbon chain may well take place during the lipoxygenase reaction. The $C_8$ alcohols act as an attractant for the phorid fly, *Megasella halterata*. This is a pest in the commercial production of mushrooms since the larvae feed on the mycelium whilst the adult fly is a vector of a mushroom disease caused by the fungus *Verticillium fungicola*. The oct-l-en-3-ol (**7.57**) is also an attractant for mites.

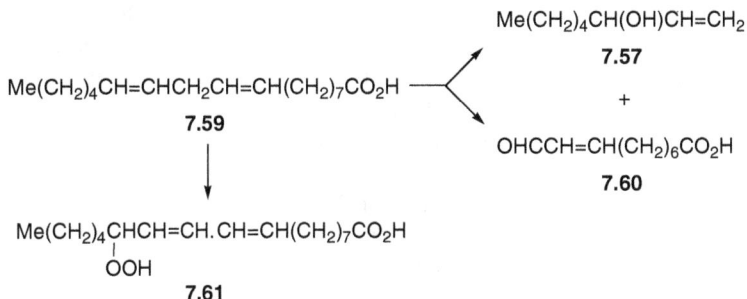

$Me(CH_2)_4CH(OH)CH=CH_2$

**7.57**

$Me(CH_2)_4CH=CHCH_2CH=CH(CH_2)_7CO_2H$

**7.59**

+

$OHCCH=CH(CH_2)_6CO_2H$

**7.60**

$Me(CH_2)_4\underset{\underset{OOH}{|}}{C}HCH=CH.CH=CH(CH_2)_7CO_2H$

**7.61**

The origin of the chlorinated 1,4-dimethoxybenzene, drosophilin A (**7.58**), has attracted some interest. This could arise by the microbial degradation of the environmental contaminant pentachlorophenol or by methylation of chloro-phenols present in the medium on which the mushrooms were grown. However, when the fungus was grown on a medium to which sodium [$^{36}$Cl]chloride had been added, labelled tetrachloro-1,4-dimethoxybenzene was formed, suggesting that it was a genuine metabolic product. Both di- and tri-chlorophenol have been detected as metabolites of other Basidiomycetes such as *Hypholoma elongatum*. This particular Basidiomycete inhabits moss in wetlands and may be the source of some of the naturally-produced chlorophenols in ecosystems.

Many other edible mushrooms produce oct-1-en-3-ol and other $C_8$ compounds together with different metabolites that give them their characteristic taste. The ratio of the $C_8$ compounds may vary. 3-Methylbutanal, 3-methylbu-tanol, benzaldehyde and phenylacetaldehyde, together with 1- and 2-phenyl-ethanol, have been detected in *Boletus edulis*.

## 7.7.2 Volatile Fungal Metabolites Containing Sulfur

The garlic Marasmius *M. alliaceus* contains a group of linear sulfur-containing metabolites such as 2,4,5,7-tetrathiooctane (**7.62**) and dimethyl polysulfides. The dipeptide γ-glutamylmarasmane (**7.63**), which contains a cysteine sulfoxide moiety, is the precursor of these garlic-like odours. A group of sesquiterpenoid alliacane metabolites has also been isolated from this fungus and are discussed in Chapter 5.

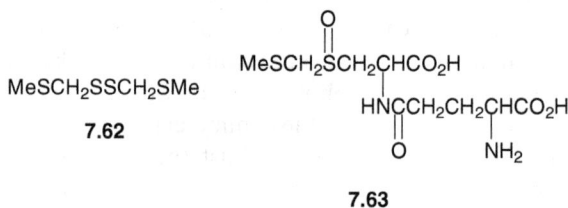

MeSCH$_2$SSCH$_2$SMe

**7.62**

MeSCH$_2$SCH$_2$CHCO$_2$H

HNCCH$_2$CH$_2$CHCO$_2$H

**7.63**

The peptide **7.63** is cleaved by a glutamyl transpeptidase and a C–S lyase to form MeSCH$_2$S.OH, which then disproportionates to give (MeSCH$_2$S)$_2$ (**7.62**). The shiitake mushroom, *Lentinus edodes*, contains a related peptide, lentinic acid (**7.65**), which undergoes similar enzymatic reactions to form lenthionine (1,2,3,5,6-pentathiepane) (**7.64**) and some other polysulfides. These contribute to the taste of the fungus. The components of the dried edible mushroom can differ significantly from the freshly harvested mushroom. The development of various pyrazines can be observed. A valuable property of the shiitake mushroom is its ability to reduce blood cholesterol levels. The active constituent is a purine de-rivative, eritadenine (lentinacin), which has the structure (2R,3R)-dihydroxy-4-(9-adenyl)butyric acid (**7.66**).

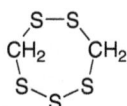

**7.64**

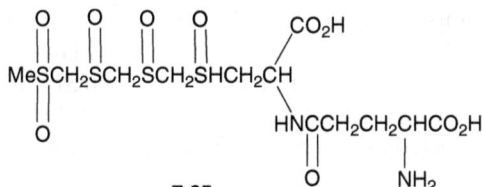

**7.65**

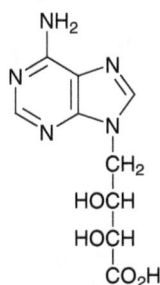

**7.66**

Several sulfur compounds have been detected as volatile constituents of truffles. 2,4-Dithiapentane is a major component of the volatile aromatic compounds of the Italian white truffle, *Tuber magnatum*. Over 120 compounds have been detected in the black Perigord truffle, *T. melanosporum*. These include dimethyl sulfide, 2-methylbutanol, 2-methylpropanal and 2-methylpropan-1-ol. The nutty and earthy flavour is attributed to anisoles and polymethoxybenzenes. Truffles also produce a volatile steroid, androst-16-en-3-one (**7.67**), which when more concentrated has an unpleasant smell. The combination of these compounds produces an odour that is a powerful animal attractant. The capacity of animals to detect the presence of underground black truffles by these substances has been evaluated by burying samples of the different compounds. The animals located the dimethyl sulfide lure as well as the black truffle flavouring.

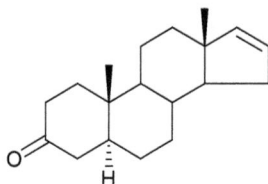

**7.67**

Dimethyl disulfide and dimethyl trisulfide, together with hydrogen sulfide and methanethiol, are major components of the odour of the stink horn, *Phallus impudicus*. Other volatile compounds include linalool, *trans*-ocimene and phenylacetaldehyde. These compounds attract flies to the stinkhorn. The spores of the stink horn stick to the fly and are then dispersed to potential sites for the organism to grow. Soil-borne and aquatic fungi also produce several volatile metabolites with a distinctive odour. Thus, some *Trichoderma* species have a characteristic coconut odour associated with the production of 6-n-pentylpyrone (**7.68**) whilst geosmin (**7.69**) and 2-methylisoborneol (**7.70**) have also been detected as odiferous fungal metabolites in some water supplies. The 6-n-pentylpyrone, which permeates the soil around the *Trichoderma* species, inhibits the growth of other fungi and is associated with the use of *T. harzianum* as a bio-control agent (see Chapter 8).

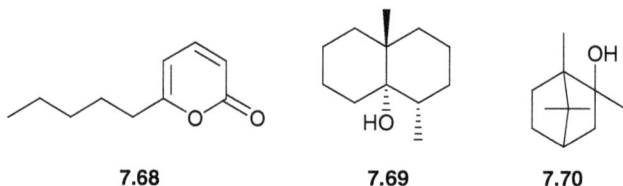

**7.68**                     **7.69**                 **7.70**

   Volatile fungal metabolites contribute to the organoleptic properties of several other foodstuffs. Surface mould ripened cheeses such as Brie and Camembert are produced using *Penicillium camembertii* and, more commonly, *P. caseicolum*. Oct-1-en-3-ol accompanied by smaller amounts of octa-1,5-dien-3-ol and 3-one are major contributors to the aroma. These organisms also have the ability to produce 2-alkanones from fatty acids and these contribute to the odour of the cheese. 2-Methylisoborneol (**7.70**) and 2-methoxy-3-isopropylpyrazine (**7.71**) have been detected in mature cultures. An unpleasant earthy flavour encountered in some aged cultures has been attributed to the excessive production of **7.71**.

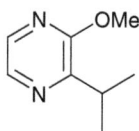

**7.71**

CHAPTER 8

# The Chemistry of Some Fungal Diseases of Plants

## 8.1 Introduction

Diseases of plants may arise from attack by fungi, bacteria, viruses, insects or parasitic plants. Insects may also act as the vectors of plant diseases and provide routes for pathogens to enter plants. The disease may be manifest by decayed or swollen roots, shrivelled or lost fruit, chlorotic or wilting leaves, necrotic lesions on the stems, leaves or fruit, excessive gum formation or the formation of cankers. In this chapter we are concerned with the chemistry of the fungal attack on plants that lead to these symptoms.

Most gardeners have experienced the unwelcome appearance of plant pathogenic fungi. The 'damping-off' of young seedlings perhaps encouraged by over-zealous watering and caused by *Pythium debarynum, P. ultimum* or *Rhizoctonia solani*, the leaf spots or grey powdery growth of *Botrytis cinerea* on lettuce leaves or strawberries or the brown patches of *Fusarium culmorum* in the lawn are just a few examples of the attack of fungi on plants. Over the last 25 years there have been noticeable changes to the countryside brought about by the ravages of Dutch elm disease (*Ceratocystis ulmi*).

Fungal diseases of crops have had a major impact on history. The devastating effect of *Phytophthora infestans* on the Irish potato crop and the resultant famine in 1845 led to a significant emigration to the United States.

Not all fungal–plant interactions are detrimental to the plant. Mycorrhizal organisms growing around the root system can facilitate the degradation of humus and the release of minerals and phosphate for uptake by the plant. The growth of certain plants, such as orchids, depends on a fungus. Endophytic organisms growing within a plant also modify the plant metabolism and can facilitate the degradation of some of the leaf constituents.

The Chemistry of Fungi
By James R. Hanson
© James R. Hanson, 2008

# 8.2    General Chemistry of Plant–Fungal Interactions

The phytopathogenic interaction between a fungus and a plant involves a combination of enzymatic and chemical steps. A generalized picture can include the following sequence of events. The initial step as the fungal spores alight on the plant is that of recognition and adhesion based on surface polysaccharides and glycoproteins. There is evidence of root exudates from plants stimulating the growth of soil fungi in the neighbourhood of the plant. In the next step, the fungus gains entry to the plant through a wound or by the action of cellulolytic enzymes. Pruning wounds are potentially a serious point of entry because the secateurs can transfer spores or mycelium into the freshly damaged plant which has not had time to heal. This entry to the plant or tree may also be mediated by an insect vector such as a bark beetle. The fungus then produces degradative enzymes that cause necrotic lesions, as well as low molecular weight toxins which damage the plant. In some cases these metabolites are host-selective or host-specific toxins. This can account for the disease specificity of particular fungal–plant interactions. The consequences of damage and infection of the trunk of a tree or stem of a plant may be manifest elsewhere as the phytotoxins are transported by the sap or there is an effect on the translocation of nutrients and water leading to chlorosis and wilts. The *in vitro* production of some phytotoxic metabolites by a plant pathogen may require the presence in the culture medium of a chemical factor found in the host plant. For example, the formation of the sesquiterpenoid phytotoxin HS-toxin by *Helminthosporium sacchari*, which is a pathogen of sugar cane, required an extract of sugar cane containing serinol (2-amino-l,3-propanediol) in the medium. This observation is particularly interesting in the light of the recent discovery that many of the genes that code for the biosynthesis of phytotoxic fungal metabolites are clustered and hence their expression could be regulated by a single compound. The major cellular targets of phytotoxins within the host plant involve the inhibition of specific enzymatic processes such as electron transport or an alteration to the permeability of cell membranes.

A plant may contain not only protective anti-fungal agents as allelochemicals but it may also respond to the attack by forming anti-fungal phytoalexins. Cell-signalling substances that stimulate the formation of phytoalexins in adjacent healthy tissue may also be released by the plant. Although the phytoalexins have a fairly general anti-fungal activity, specific plant pathogens have developed the ability to metabolize and detoxify the phytoalexins produced by their host. The specific plant pathogen then has the double advantage of a competition-free environment and a weakened plant. Other fungi may metabolize the constitutive allelochemicals. Conversely, some resistant plants also have the ability to metabolize the phytotoxins produced by the invasive fungus.

Fungi that are associated with plants do not act in isolation. Some produce anti-fungal and anti-bacterial agents to ensure their ecological niche. There are also interactions with insect vectors. It does not take much imagination to realize that there is a great deal of chemistry in fungal phytotoxicity. In the following we illustrate these facets of chemical microbiology with some specific examples from a range of phytotoxic effects.

# 8.3 Chemistry of some Leaf-spot Diseases

## 8.3.1 *Botrytis cinerea*

The fungus *Botrytis cinerea* is a grey powdery mould that is a serious pathogen of several commercial crops, including lettuce, grapes, strawberries and tobacco. It has been found on over 200 plants. It is one of several phytopathogens that produce brown necrotic lesions on leaves and fruit. Several other phytopathogens belong to the same genus, including *B. allii* and *B. squamosa*, which are found on onions, *B. fabae*, which is found on beans, and *B. gladioli* and *B. tulipae*, which are found on gladioli and tulips. The genus *Botrytis* belongs to the order Moniales of the Fungi Imperfecti. The name Botrytis is derived from the Greek βοτρυξ, meaning a bunch of grapes, which describes the appearance of the conidia on the conidiophores. Because of the economic losses arising from a *Botrytis* infection, there has been a great deal of work on the phytopathology, microbiology and chemistry of this fungus.

The conidia are disseminated by the wind and can persist in the soil or on plant debris. Some insects can facilitate their transmission. For example, the larvae of the moth *Lobesia botrana*, which has been found on grapes, carry the infective pathogen. The larvae are responsible for the distribution and for creating a point of entry to the plant by damaging the cuticle. The larvae increase the supply of nutrients to the fungus not just as a result of damage to the plant but also by their secretions.

Several distinctive stages in the adhesion of the conidia to the plant surface have been recognized. These are based on a sheath of glycoprotein surrounding the conidia. Germination on the surface of a plant requires the presence of nutrients such as sugar, phosphates and a nitrogen source. These may be provided by the damaged plant or even by pollen deposited on the plant surface. Epiphytic organisms such as bacteria and yeasts can discourage the germination of conidia. Antibiotic formation, *e.g.* pyrrolnitrin by a *Pseudomonas* species, may influence this. Some fruit such as strawberries can suffer a latent infection in which *Botrytis cinerea* has penetrated the immature fruit but the mycelium does not develop until the fruit ripens. In this case the germination of the conidia may be influenced by the concentration of the fruit acids, malic and tartaric acid, which tend to diminish as the fruit ripens. There is a favourable temperature (19–26°C) and humidity for the development of the fungus.

As the fungus develops it produces necrotic lesions on the leaf and the fruit surfaces. These are the result of the action of cellulolytic enzymes and pectinases. The latter will hydrolyse the methoxyl groups of pectin, releasing polygalacturonic acid, which is then further hydrolysed. The breaking of the pectin chain leads to lysis of the cell. *Botrytis cinerea* also produces laccase enzymes, which oxidize phenolic substances, and other oxidases, which produce reactive oxygen species that participate in lipid peroxidation and degradation. The lesions on the grapes allow water to evaporate and the grape to become dehydrated with a higher sugar concentration. Fermentation of these grapes can produce a sweet 'spatlese' wine. The phenolic oxidation products contribute to the golden-yellow colour of wines produced by botrytized grapes.

The fungus produces two series of low molecular weight phytotoxins, the sesquiterpenoid botryanes such as botrydial (**8.1**) and dihydrobotrydial (**8.2**) and the polyketide botcinolides, *e.g.* **8.3**. The most phytotoxic of the botryanes is botrydial. Not only is this produced in laboratory cultures of the fungus but it has also been detected in plants infected by *B. cinerea*. Typical of several phytotoxic agents, it contains a reactive dialdehyde that could react in the cell with various amino groups. The biosynthesis of the botryane sesquiterpenoids has been thoroughly investigated and has been described in Chapter 5. Interference with this biosynthesis provides a potential means of controlling the development of a *Botrytis* infection. Botrydial (**8.1**) is subsequently metabolized to the less toxic dihydrobotrydial (**8.2**) and then to a series of degradation products. When the botrydial concentration reaches a particular level, the growth of the fungus is restricted until its further metabolism reduces the levels of botrydial. Analogues of botrydial and its immediate precursors can mimic this effect at an early stage in the infection and provide a novel way of selectively controlling the development of this pathogen.

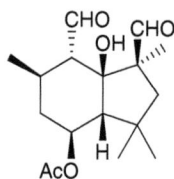

8.1

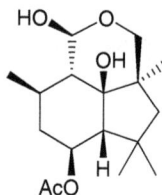

8.2

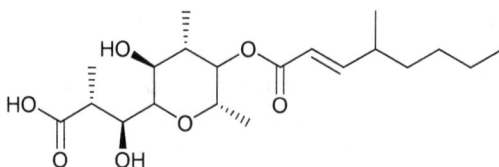

8.3

A plant not only contains compounds that prevent the growth of phytopathogenic organisms but, on being attacked, the production of phytoalexins is stimulated. An example of a volatile natural anti-fungal agent is the leaf aldehyde, hex-2-en-l-al and its double bond isomer. The presence of curcubitacins in cucumber protects these plants against *Botrytis*. Some examples of phytoalexins produced by plants as a consequence of attack by *B. cinerea* are capsidiol (**8.4**) (from peppers, *Capsicum frutescens*), wyerone (**8.5**) produced by the bean *Vicia fabae*, luteone (**8.6**), produced by the lupin *Lupinus luteus*, and the stilbene resveratrol (**8.7**) produced by the grape *Vitis vinifera*. These afford general protection against fungal attack. However, they are metabolized by strains of *B. cinerea* to less toxic metabolites. The strains of *B. cinerea* that can degrade these are highly pathogenic to the plant. The strains that are unable to achieve this are not pathogenic. Thus capsidiol may be oxidized to capsenone (**8.8**). Wyerone is converted into the hexahydro derivative. The alkene of the prenylated isoflavone luteone is epoxidized and the epoxide is then modified. Resveratrol is oxidized to a stilbene dimer (**8.9**). These detoxification mechanisms are part of the biological warfare that is a part of the plant–fungal competition.

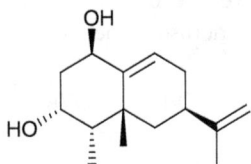

**8.4**

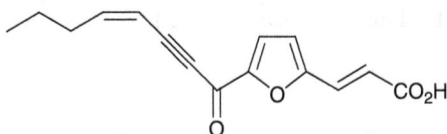

**8.5**

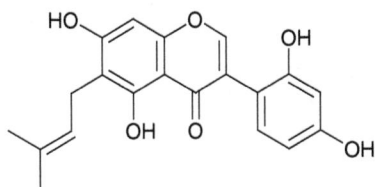

**8.6**

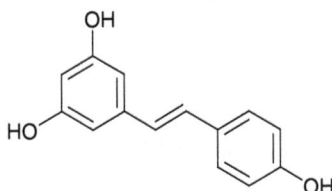

**8.7**

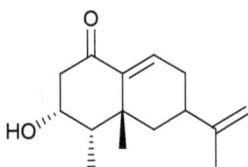

**8.8**

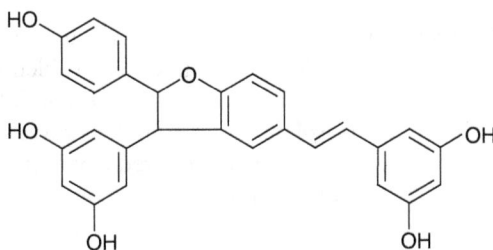

**8.9**

## 8.3.2  *Alternaria* Leaf-spot Diseases

*Alternaria* species are associated with several leaf-spot diseases of plants. Alternaric acid (**8.10**), produced by *A. solani*, which is the cause of 'early blight' on potatoes and tomatoes, has a marked phytotoxic activity and may be responsible for many of the symptoms of the plant disease caused by the fungus. It also shows a specific anti-fungal activity. The branched chain structure is biosynthesized from nine acetate units and three $C_1$ units derived from formate. The solanopyrones are another group of phytotoxic metabolites of *A. solani*. Solanopyrone A (**8.11**) induced necrotic lesions on the leaves of potatoes typical of this fungal infection. The structures of the solanopyrones were established by a combination of chemical and spectroscopic methods. These compounds have also been isolated as the phytotoxic metabolites of *Ascochyte rehiei*, which causes a disease of chick pea. The solanopyrones are biosynthesized by a route that involves an interesting Diels–Alder reaction. The pyrone is formed from eight acetate units and two methyl groups from methionine. Hydroxylation and oxidation affords the achiral linear triene prosolanopyrone III (**8.12**). This triene is the substrate for solanopyrone synthase to give the chiral solanopyrone A (**8.11**). This synthase was the first (1995) Diels–Alderase to be purified.

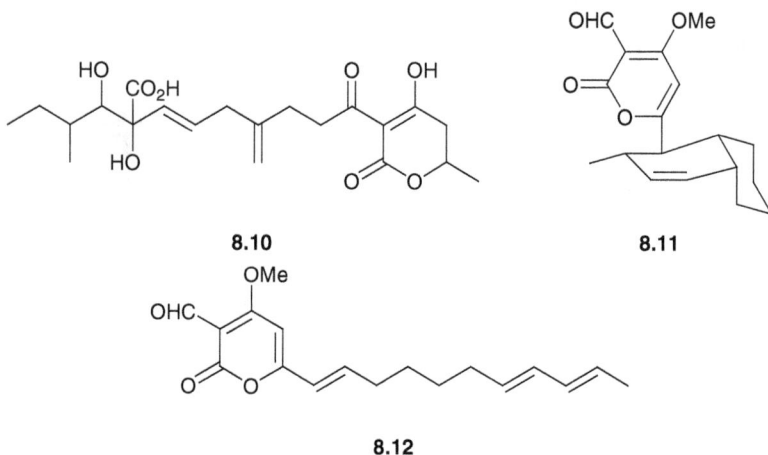

**8.10**

**8.11**

**8.12**

Another widespread species, *A. tenuis*, flourishes as a saprophyte on dead or weakened plants. This fungus can cause problems in paper manufacture because of its ability to discolour wood pulp. Examination of the mycelium of a strain of *A. tenuis* that was isolated from tomatoes revealed alternariol (**8.13**) and its methyl ether. Alternariol is biosynthesized from nine acetate units. This compound is related to the dienone botrallin (**8.14**), from *Botrytis allii*. Extraction of the broth of an *A. tenuis* fermentation afforded a group of closely related heptaketides. Their structures, which were primarily established by X-ray crystallography, can be related to altenusin (**8.16**). Oxidation of altenusin affords the dienone dehydroaltenusin (**8.15**), which was in turn related to altenuene. Cleavage of the catechol ring of altenusin to a dicarboxylic acid followed by lactonization gave altenuic acid II (**8.17**). Altenuisol lacks the C-methyl group of dehydroaltenuisin.

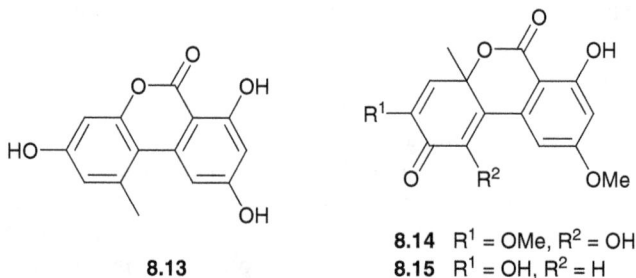

**8.13**

**8.14** $R^1$ = OMe, $R^2$ = OH
**8.15** $R^1$ = OH, $R^2$ = H

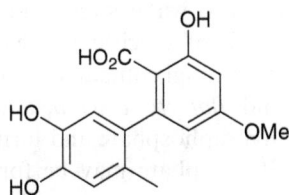

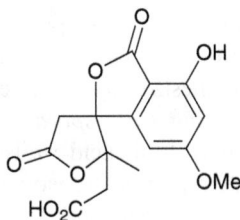

8.16                                    8.17

The tetramic acid tenuazonic acid (**8.18**), which was also obtained as a meta-bolite of *A. tenuis*, has been isolated as the major toxin of *A. alternata* and from *Pyricularia oryzae*. The latter is the causal organism of a rice blast disease. Tenuazonic acid produces a stunting effect on the growth of rice seedlings. The structure of tenuazonic acid was established by chemical degradation. On treatment with acid, tenuazonic acid gave acetic acid and a deacetyl compound that behaved as a β-diketone. Ozonolysis and alkaline degradation gave iso-leucine and acetic acid. The structure was confirmed by synthesis. The medium on which *A. alternata* is grown has a significant bearing on the production of its metabolites.

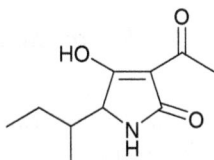

8.18

The enolic tautomers of tenuazonic acid readily form copper, magnesium and calcium complexes and their formation has been associated with its biological activity. The fungus *Phoma sorghina*, which is found growing on sorghum and millet in sub-Saharan Africa, has been associated with a haematological disease of man known as Onyalai. Magnesium and calcium tenuazonates have been isolated as the toxic constituents of this fungus.


### 8.3.3   *Cercospora* Leaf-spot Diseases

*Cercospora* species are responsible for various leaf-spot diseases. *C. rosicola* produces small brown lesions surrounded by a red-violet ring on rose leaves. The common black spot is caused by a different fungus, *Diplocarpon rosae*. A severe *Cercospora* infection can cause loss of the leaf. The phytotoxic fungus *C. rosicola* has been shown to produce the plant hormone abscisic acid (**8.19**), which is sometimes known as the dormancy hormone. This plant hormone induces various adaptive responses in plants to water-stress and the onset of low temperatures. Studies on the biosynthesis of abscisic acid have suggested that it is biosynthesized by a different route in fungi compared to higher plants. In fungi, including *Cercospora* species and *Botrytis cinerea*, abscisic acid is formed from mevalonic acid *via* isopentenyl diphosphate and farnesyl diphosphate. In higher plants the isopentenyl diphosphate may be formed *via* the l-deoxy-D-xylulose 5-phosphate pathway. There are then two pathways, a direct pathway involving the cyclization of farnesyl diphosphate to an ionylidienethanol (**8.20**) and oxidation to abscisic acid or a carotenoid pathway that involves the formation and oxidation of (9*Z*)-carotenoids. The balance between these two pathways is not clear although it appears that fungi follow the direct pathway.

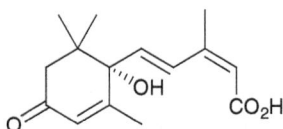

**8.19**

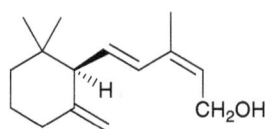

**8.20**

### 8.3.4   Diseases Caused by *Colletotrichum* Species

Anthracnoses are plant diseases in which there is a characteristic dark oval necrotic spot or even a lesion on the leaf or fruit of a plant. Several economic crops, such as grapes, strawberries, beans, peppers and tobacco, are susceptible to these diseases, which are often caused by *Colletotrichum* (*Glomerella*) species of fungi. The metabolites that are responsible for the symptoms of the diseases have been isolated from several of these organisms. A widespread fungus is *Colletotrichum gloeosporioides* (*Glomerella cingulata*), which produces a range of metabolites of different structural types. Phytotoxicity has been associated with some diketopiperazines, *e.g.* **8.21**, and with cyclic peptides that chelate metal ions as siderophores. Other metabolites such as gloeosporone (**8.22**) and the mycosporins, *e.g.* **8.23**, affect the development of fungi. An endophytic strain has been shown to produce the unusual colletotric acid (**8.24**).

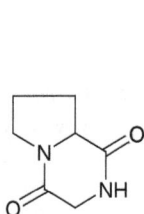

**8.21**

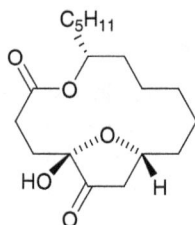

**8.22**

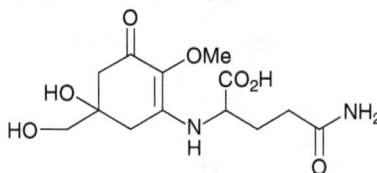

**8.23**

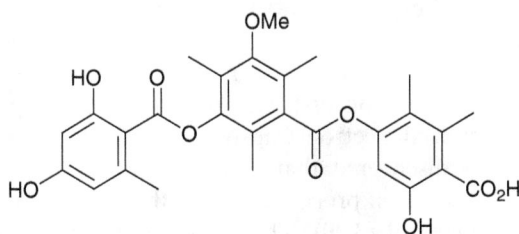

**8.24**

Tobacco anthracnose disease is caused by *C. nicotiana*. The phytotoxic colletotrichins, *e.g.* **8.25**, which are produced by this fungus, are meroterpenoids. Their biosynthetic labelling pattern from acetate is commensurate with a mixed norditerpenoid–polyketide origin. When put on tobacco leaves, a dilute solution of the metabolite colletopyrone (**8.26**) reproduces the necrotic lesions characteristic of the disease. The fungus *C. capsici*, which is a pathogen on peppers, produces the phytotoxic bislactones colletodiol (**8.27**) and colletoketol, which are described in Chapter 4.

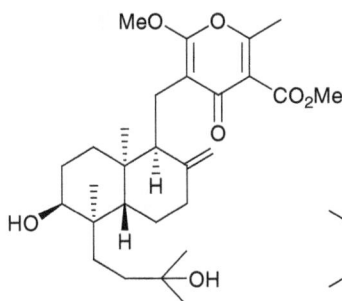

**8.25**

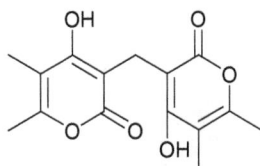

**8.26**

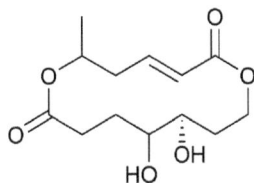

**8.27**

## 8.4 Fungal Diseases of the Gramineae

Wheat, barley, oats and rice provide a staple food for a substantial portion of the world's population. Consequently, their fungal diseases, and their means of defence against them, have attracted considerable attention.

Fungi of the genus *Helminthosporium* are responsible for several leaf spot and other diseases of cereals and grasses. After the initial work on their metabolites had been reported, many of these organisms were reclassified as *Bipolaris* species. The perfect stage of some of these fungi is *Cochliobolus* and hence there can be confusion as to the source of a metabolite.

The pigmentation associated with some of these fungi has been associated with anthraquinone formation and has been described in Chapter 7. However, the phytotoxicity has been attributed at least in part to the formation of sesquiterpenes; helminthosporal (**8.28**) and helminthosporol (**8.29**) were first isolated from *H. sativum*, now known as *Bipolaris sorokinianum*, and their structures were established by de Mayo in 1962 (see Chapter 5). Helminthosporal inhibited the germination of barley seedlings and it appeared to be an inhibitor of mitochondrial electron transport and oxidative phosphorylation. Subsequently, it was discovered that helminthosporal and helminthosporol were artefacts of the isolation process and that they were formed from acetals such as prehelminthosporol (**8.30**). This compound inhibits wheat coleoptile growth. Several related phytotoxic metabolites have been isolated more recently, including sorokinianin (**8.31**), which possesses an additional three-carbon unit, and the cyclic amine victoxinine (**8.32**). The latter was first isolated as a phytotoxin from the culture filtrates of *H. victoriae*, the causal agent of 'Victoria blight' of oats.

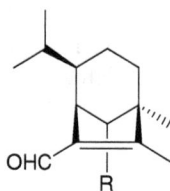

**8.28**  R = CHO
**8.29**  R = CH₂OH

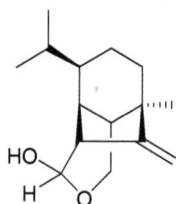

**8.30**

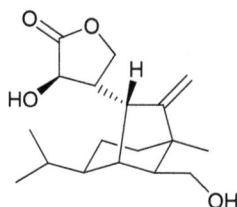

**8.31**

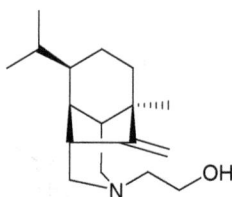

**8.32**

These fungi are also pathogens on some weeds and have been examined in the context of developing new herbicides. Thus, prehelminthosporol has been isolated from a *Bipolaris* species that was a pathogen on a serious tropical weed known as Johnson grass (*Sorghum halepense*).

The fungus *H. carborum* selectively attacks certain types of maize. It has been found to produce a host-specific toxin, HC-toxin, which has been shown to be a cyclic tetrapeptide, cyclo-[(2-amino-9,10-epoxy-8-oxodecanoyl)-alanyl-alanyl-prolyl]. This is one of several cyclic tetrapeptides that affect higher plants.

These crops possess constitutive allelochemicals that provide protection against attack by fungal pathogens. Graminaceous plants contain the hydroxamic acids 1,4-benzoxazin-3(4*H*)-ones DIMBOA (**8.33**) and DIBOA (**8.34**) and related benzoxazoles **8.35** and **8.36** which are stabilized as their glycosides. The roots of oats contain the triterpene avenacin (**8.37**) as an allelochemical. Various strains of the root-infecting fungus *Gaeumannomyces graminis* are serious pathogens of cereal crops. These strains have different host specificities, which are reflected in their interaction with plant defensive chemicals. *G. graminis* var. *tritice* causes 'take-all' of wheat and barley. *G. graminis* var. *graminis* is a pathogen of wheat and rice whilst *G. graminis* var. *avenae* is the cause of 'take-all' in oats. There is some mutual antagonism between these varieties. The wheat pathogens *G. graminis* var. *graminis* and var. *tritici* degrade and detoxify the cyclic hydroxamic acids and the benzoxazoles (BOA and MBOA) to *N*-(2-hydroxyphenyl)-malonamic acid and *N*-(2-hydroxy-4-methoxyphenyl)malonamic acids, respectively. The ability of the *Gaeumannomyces* isolates to cause the symptoms of root rot in wheat parallels their ability to detoxify the wheat allelochemicals. *G. graminis* var. *avenae* has an enzyme that removes the glucose units from the triterpenoid saponins avenacins A and B, and thus detoxifies them.

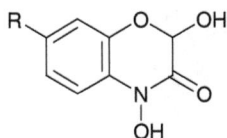

**8.33** R = OMe
**8.34** R = H

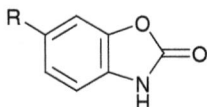

**8.35** R = OMe
**8.36** R = H

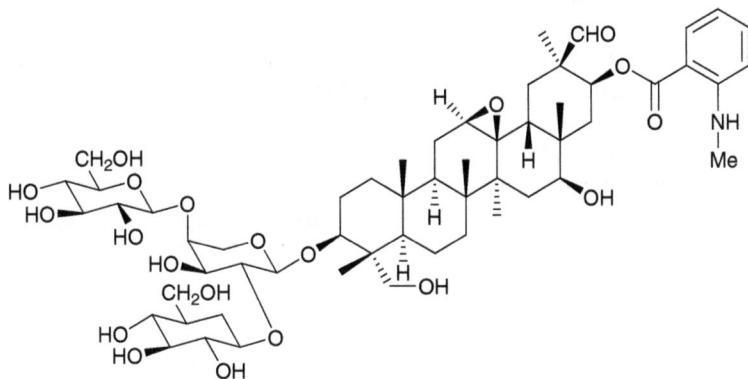

**8.37**

# 8.5 Root-infecting Fungi

The genus *Fusarium* consists of very common soil microfungi, including numerous plant pathogens. The nomenclature of the Fusaria has undergone various revisions. In the authoritative book by Booth on the genus *Fusarium* it is noted that many of the names are merely host epithets and represent a new record of a fungus on a particular host rather than a new species.

*Fusarium monoliforme, F. graminearum, F. avenaceum* and *F. culmorum* are pathogens of the Gramineae whilst *F. solani* has been isolated from a wide range of plants in which it causes root rots, particularly of various members of the Solanaceae and beans. *F. oxysporum* var. *cubense* produces a devastating wilt of bananas.

The pigments of the Fusaria have been described in Chapter 7. Many of the characteristic phytotoxic metabolites are the sesquiterpenoid trichothecenes (**8.38**) (see Chapter 5). The more highly hydroxylated members have considerable mammalian toxicity as well as phytotoxicity. Trichothecenes have been identified as metabolites of species from ten of the twelve sections of the genus *Fusarium* as classified by Booth. The trichothecenes occur with various combinations of oxygen substituents at positions 3, 4, 7, 8 and 15. Several trichothecenes contain macrocyclic esters linking C-4 and C-15. These are known as the roridins and verrucarins.

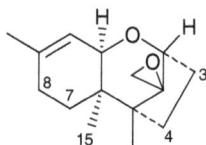

**8.38**

Since many trichothecene-producing fungi are plant pathogens, the phytotoxicity of these metabolites has been investigated. Trichothecin (**8.39**), diacetoxyscirpenol (**8.40**) and T-2 toxin (**8.41**) are quite common examples and produce scorching of the foliage of pea seedlings at a concentration of 1 μg mL$^{-1}$. At a concentration of 10 μg mL$^{-1}$, diacetoxyscirpenol severely reduced the stem elongation of pea seedlings and many of the test plants died. There is other circumstantial evidence that implicates trichothecenes in phytotoxicity. For example, a strain of *F. tricinctum*, which was isolated from turf that had undergone a yellow wilt, produced a good yield of T-2 toxin in the laboratory. Although these compounds are phytotoxic, they are better known as mycotoxins and are discussed in Chapter 9.

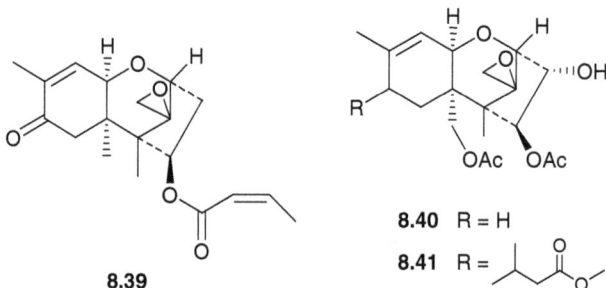

**8.39**

**8.40**  R = H

**8.41**  R =

The fungus *Penicillium gladioli* is a common cause of the Penicillium rot on gladiolus corms. The fungus produces a strongly anti-fungal and weakly anti-bacterial substance, gladiolic acid (**8.42**), which allows it to develop in a potentially competitive spoilage situation. Under somewhat different fermentation conditions the fungus produces dihydrogladiolic acid (**8.43**). The dialdehyde structure of gladiolic acid is interesting because it exists in tautomeric equilibrium with the lactol form, a feature that was studied in 1953 in an early application of infrared spectroscopy to structural analysis. The structural and biosynthetic studies on gladiolic acid were discussed in Chapter 4.

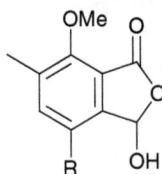

**8.42** R = CHO
**8.43** R = CH$_2$OH

# 8.6   Some Fungal Diseases of Trees

## 8.6.1   Dutch Elm Disease

The relationship between fungi and insects is complex. The bark beetle, *Scolytus multistriatus*, is the vector of Dutch elm disease, which is caused by the fungus *Ceratocystis ulmi*. The elm produces catechin 7-xyloside and a triterpenoid (lupeyl cerotate), which are feeding attractants for the bark beetle. The beetle brings with it the spores of the fungus. This develops within the galleries bored by the beetle and produces two types of toxins, a glycoprotein and some phenolic metabolites. The C$_{10}$ acid 2,4-dihydroxy-6(l-hydroxyacetonyl)benzoic acid (**8.44** R = H, OH) together with the 6-acetonyl (**8.44** R = H$_2$) and 6-pyruvyl (**8.44** R = O) analogues exist in part as their lactol tautomers **8.45**. They are produced more rapidly and in greater yield by the 'fluffy' aggressive strains of *C. ulmi*. These acids had been obtained previously from *Penicillium brevicompactum*. The 6-acetonyl compound, which is moderately phytotoxic, has also been isolated from the plant pathogens *Alternaria kikuchiana* and *Pyricularia oryzae*, whilst the related isocoumarin diaporthin (**8.46**) is the phytotoxin of *Endothia parasitica*, which is a pathogen of chestnuts.

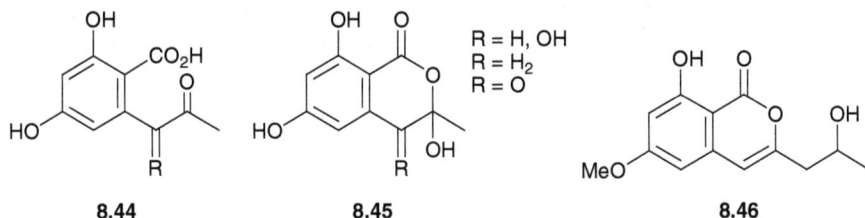

R = H, OH
R = H$_2$
R = O

**8.44**          **8.45**          **8.46**

The phenolic and glycoprotein metabolites of *C. ulmi* cause the wilting and death of the tree. However, the beetles are deterred from feeding and using as brood trees those trees that have already been infected by another fungus, *Phomopsis oblonga*. This fungus is frequently found growing on the outer bark of elm trees and it can invade the phloem of stressed trees that are already infected by *C. ulmi*. It produces deterrent compounds, including 5-methylmellein (**8.47**) and dihydropyrones, the phomopsolides A (**8.48**) and B. The phomopsolides have anti-boring and anti-feeding activity against the elm-bark beetle. The use of traps containing the beetle aggregation pheromones (δ-multistriatin, 4-methyl-3-heptanol and α-cubebene) enables an estimate to be made of the number of beetles in an area and hence the potential for infection of elms by *C. ulmi*. The phomopsolides have also been isolated from a *Penicillium* species that was found growing as an endosymbiont in the inner bark of the Pacific yew, *Taxus brevifolia*. It is possible that these compounds may confer a wider protective role against bark-boring beetles.

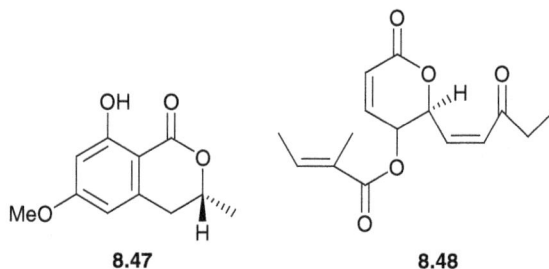

8.47          8.48

## 8.6.2    Eutypa Dieback

Various fungal diseases of woody plants enter the plant through wounds caused by pruning. Not only can these infections cause necrosis of the adjacent woody tissue but their consequences can spread to new shoots and to more distant parts of the plant. In the case of fruit bushes and trees this can affect the yield. Eutypa dieback caused by the fungus *Eutypa lata* is a canker disease that affects, amongst other plants, almonds, apricots, grapes, cherries, peaches and walnuts. It is often spread by pruning. Infected plants eventually die. The infection of vines by this fungus leads to a serious loss of wine production. The severity of the disease is associated with the production of a toxin, eutypine (**8.49**), and several related metabolites, including the benzofuran **8.50** and the chromene eulatachromene (**8.51**). The related benzodihydrofuran is fomannoxin, which is produced by *Fomes (Heterobasidium) annosus*, a fungus that is a serious white rot of conifers.

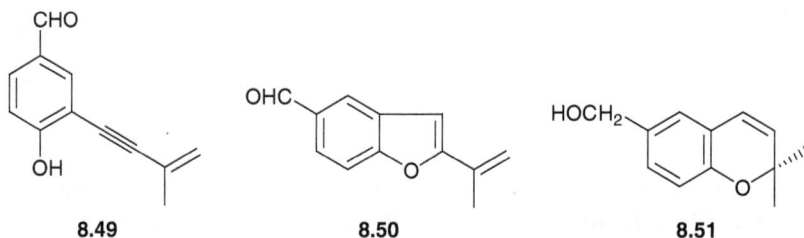

8.49                    8.50                    8.51

Eutapine is thought to be transported in the ascending sap to the leaves and flowers where it produces its effect by uncoupling oxidative phosphorylation. Some grapevines can detoxify eutapine by an enzymatic reduction of the aldehyde to the primary alcohol eutypinol which is inactive. There is a relationship between the susceptibility of grape cultivars to Eutypa dieback and their ability to reduce eutypine. Efforts have been made to enhance this protective effect. Various other strains of the fungus produce some biologically active highly oxygenated cyclohexane epoxides such as **8.52**.

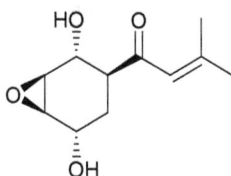

**8.52**

### 8.6.3 *Armillaria mellea*

A common root-rot of trees is caused by the honey-fungus *Armillaria mellea*. Infection by this Basidiomycete can lead to the death of trees such as apples, flowering cherries, willows, birch and walnuts. The fungus spreads between trees by underground rhizomorphs, hence the other name of the fungus: the boot-lace fungus. The fruiting body, which appears near an infected tree, is honey coloured. The mycelium can spread in the infected tree as a white sheet between the bark and the wood of the tree. This blocks the vascular system and prevents the transport of water and nutrients to the leaves, which then rapidly wilt and die. The fungus produces several sesquiterpenoid phytotoxic metabolites, exemplified by armillyl orsellinate (**8.53**) and melleolide (**8.54**) (see Chapter 5). These compounds also have antibiotic properties that facilitate the survival of the fungus in a competitive environment.

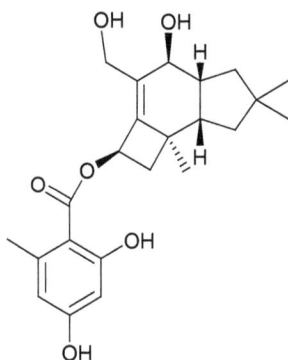

**8.53**

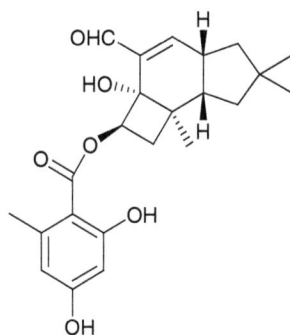

**8.54**

### 8.6.4    *Phytophthora cinnamomi*

The fungus *Phytophthora cinnamomi* causes a root and stem-base disease of a wide-range of coniferous and broad-leaved trees. For example, it is associated with the decline of several species such as the beech and the Spanish oak in particular environments. It secretes an elicitor protein, β-cinnamomin, which elicits plant defence mechanisms and in sensitive species produces plant cell necrosis. This protein acts as a sterol carrier protein and so it affects cellular membranes. A related organism, *P. ramorum*, is the causal agent of a disease known as 'Sudden Oak Death'. It has been found in this country, along with *P. kernoviae*, on some oak and beech trees as well as on some non-native shrubs such as viburnum and rhododendron.

### 8.6.5    Silver-leaf Disease

The silver-leaf disease of apples, plums and other fruit trees is caused by the fungus *Chondrostereum purpureum* (*Stereum purpureum*). This Basidiomycete produces a toxic protein, an endo-polygalacturonase and some sesquiterpenes which produce the disease symptoms. The sesquiterpenes possess the sterpurene skeleton, which is derived by the same cyclization that leads to the other ses-quiterpenes found in the Basidiomycetes. The biological activity of the fungus has been associated with sterpuric acid (**8.55**) and some dihydroxysterpurane metabolites, exemplified by **8.56**. The fungus is quite selective in its action and it has been used as a biocontrol agent (BioChon) in woody vegetation, where it prevents further stump sprouting when hardwoods have been harvested.

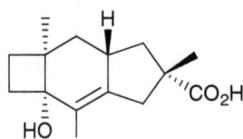   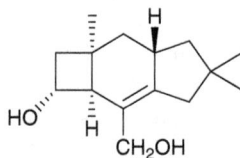

       **8.55**                                            **8.56**

### 8.6.6    *Nectria galligena* Canker

The phytopathogenic fungus *Nectria galligena* causes a canker on many trees, including apples. The asexual stage of the fungus is known as *Cylindrocarpon heteronemum*, and several other *Nectria* species are found on trees and cause damage. The fungus *N. galligena* produces a series of prenylated phenols exemplified by colletorin B (**8.57**) colletochlorin B (**8.58**) and ilicicolin E (**8.59**) as well as α,β-dehydrocurvularin (**8.60**). These have inhibitory effects on seed germination and plant development as well as producing damage to plant tissues.

**8.57** R = H
**8.58** R = Cl

**8.59**

**8.60**

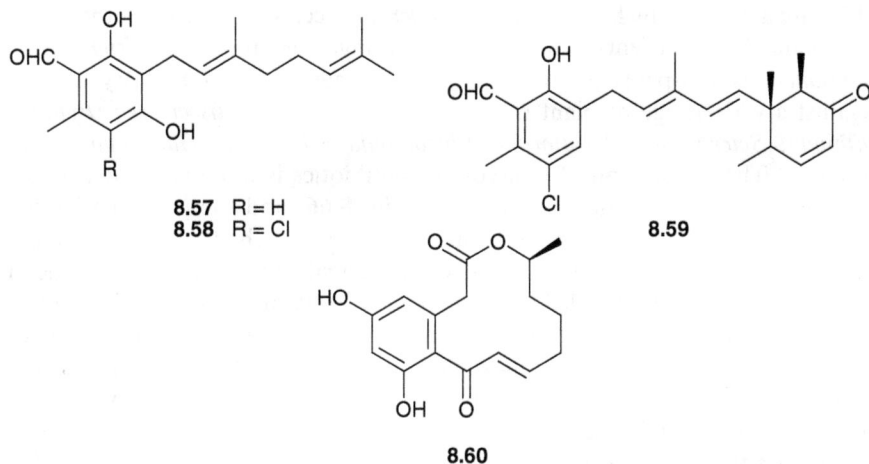

## 8.6.7 Canker Diseases of Cypress

Several species of *Seiridium*, including *S. cardinale*, *S. cupressi* and *S. unicorne*, are associated with the canker diseases of cypress, *Cupressus sempervirens*. The phytotoxic metabolites include the butenolides seiridin (**8.61**) and isoseiridin (**8.62**) and their 7′-hydroxy relatives as well as the sesquiterpenes seiricardine A (**8.63**) and B (**8.64**). Solutions of these sesquiterpenes when applied to leaves produce chlorosis, browning and death of the leaves.

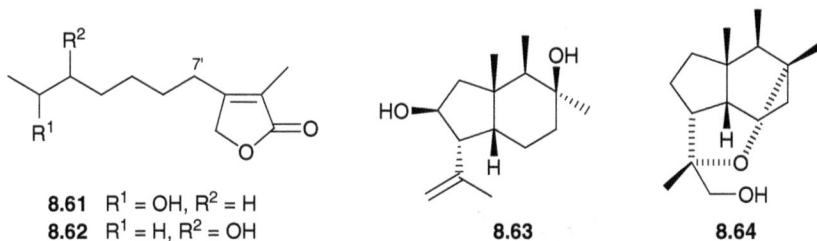

**8.61** R$^1$ = OH, R$^2$ = H
**8.62** R$^1$ = H, R$^2$ = OH

**8.63**

**8.64**

## 8.7 *Trichoderma* Species as Anti-fungal Agents

*Trichoderma* species are widespread soil-borne fungi that display significant antagonism against various other fungi, including many plant pathogens. However, *Trichoderma* species themselves display a relatively low phytotoxic potential. There are three stages in the antagonistic activity of these organisms. In the first stage some *Trichoderma* species produce volatile antibiotics that permeate through the soil and inhibit the germination of other fungi. In the second stage they produce non-volatile anti-fungal metabolites whilst in the third stage they produce extracellular enzyme systems with cellulolytic and chitinase activity, which allow the *Trichoderma* species to penetrate other fungi to obtain their nutrients. The volatile antifungal metabolite of *Trichoderma harzianum* is 6-n-pentylpyrone (**8.65**), which has potent inhibitory activity against a wide range of plant pathogens, including *Botrytis cinerea, Pythium ultimum, Sclerotinia sclerotiorum, Rhizoctonia solani* and *Gaeumannomyces graminis*. The production of non-volatile antibiotics is more strain dependent but they include the butenolide harzianolide (**8.66**) and the pyridone harzianopyridone (**8.67**). A cyclic peptide mixture, known as the trichorzianines, has also been isolated and shown to possess anti-fungal activity. The wide range of its anti-fungal activity has led to *T. harzianum* being marketed as a biocontrol agent (Rootshield® or Plantshield®). The successful application of this fungus as a bio-control agent rests upon the diversity of anti-fungal agents and extracellular enzyme systems that it produces. The formation of this wide range of antagonistic metabolites minimizes the development of resistant organisms and allows the *Trichoderma* to maintain its competitive advantage.

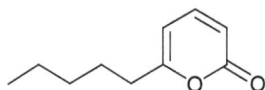

**8.65**

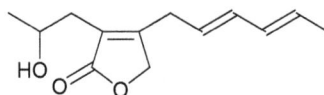

**8.66**

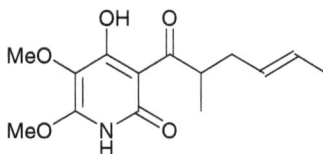

**8.67**

## 8.8   Fungal Diseases of Plants and Global Warming

The fungal diseases of plants are likely to be affected by global warming in several ways. Warmer summers are likely to favour the spread of diseases such as Dutch elm disease that rely on an insect vector, whilst drier summers, by increasing the extent of water stress, favour those organisms that attack weakened plants. The spread of phytophagous insects may widen the impact of fungal diseases of plants. Warmer and wetter winters are likely to favour soil-borne organisms that can over-winter provided the conditions are mild.

# CHAPTER 9
# *Mycotoxins*

## 9.1 Introduction

Most of us are aware of the relatively rapid effects of bacterial infections but, apart from the ingestion of poisonous mushrooms, we are probably less aware of the insidious longer term but equally devastating effects of fungal toxins. Bacterial diseases often arise as a consequence of the replication of bacteria and the resultant toxins that are produced within the body. In contrast, the deleterious impact of fungi on man may arise from the ingestion of the mature fungus and the metabolites that it has already produced. Whereas many bacterial diseases are spread by the growth of organisms from contaminated meat and water or droplet infection, the effects of fungal toxins can arise from metabolites that are present in contaminated cereals, nuts and fruit. The common mycoses, such as 'Athlete's Foot', occur on the surface of the skin or in oral cavities and in the lungs exposed to the air (Farmer's Lung). The commonest dermatophytes are *Trichophyton rubrum* and *T. mentagrophytes*. Dandruff in the hair is a consequence of the growth of another fungus, *Malassezia globosa*. Farmer's Lung is an allergic reaction to mouldy hay contaminated with *Aspergillus* species.

## 9.2 Ergotism

The oldest disease to be attributed to mycotoxins is ergotism. This disease is caused by a filamentous fungus, *Claviceps purpurea*, which is found in the ears of grain, particularly rye where it forms black sclerotia known as ergot. This fungal material found its way into rye bread and the metabolites caused epidemics of mass poisoning known in the Middle Ages as St Anthony's Fire. Diseases with symptoms resembling ergotism were even recorded by the Ancient Greeks. The use of rye to make bread, particularly in France in the Middle Ages, led to the incidence of ergotism. The burning sensation and subsequent gangrene led to the name of Holy Fire. The disease was seen as a

The Chemistry of Fungi
By James R. Hanson
© James R. Hanson, 2008

punishment from God. The name St Anthony's Fire came from a series of monastic hospices set up in France in the 11th and 12th century to help those afflicted with this disease.

Ergotism involves gangrenous effects on the extremities arising from a restriction to the circulation and to neurological effects that produce convulsive epileptic-like seizures. Another vasoconstrictive effect produced uterine contractions. This led to the induction of labour in childbirth and the use of ergot in abortions. Another member of the same genus, *C. paspali* which grows on some grass pastures, produces paspalum staggers in cattle.

Chemical studies of ergot lasting well over 100 years led to the isolation of the ergot alkaloids exemplified by ergotamine (**9.1**) and the yellow ergochrome pigments such as secalonic acid A (**9.4**). The ergot alkaloids are mostly peptide derivatives of lysergic acid (**9.2**). Members of the ergot alkaloids produce the peripheral vasoconstriction, gangrene and necrosis of the extremities whilst others produce the neurological disturbances that led to the hallucinations and seizures characteristic of St Anthony's fire. Lysergic acid diethylamide (LSD, **9.3**) is a derivative of lysergic acid (**9.2**), which is part of ergotamine. The ergochrome secalonic acid A (**9.4**) produces liver necrosis. Sporadic outbreaks of ergotism occurred well into the 20th century. The effect on uterine blood vessels led to the development of ergometrine for use in the treatment of haemorrhages in child birth. Several ergot derivatives are used in medicine, including bromocriptine, cabergoline, ergotamine, ergonovine, methylsergide and pergolide.

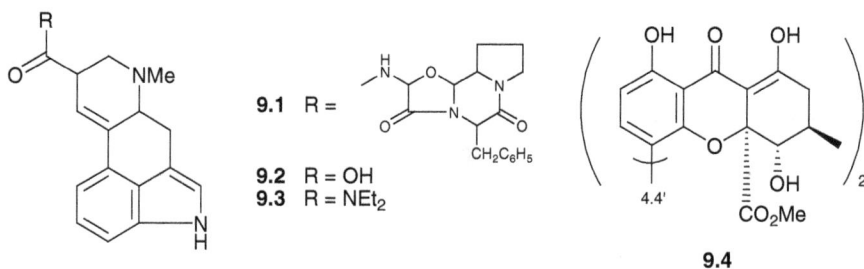

**9.1** R =

**9.2** R = OH
**9.3** R = NEt$_2$

**9.4**

## 9.3 Trichothecenes as Mycotoxins

The trichothecenes have the underlying sesquiterpenoid skeleton (**9.5**). Their chemistry and biosynthesis was discussed in Chapter 5. Alimentary toxic aleukia is a serious mycotoxicosis that is caused by fungi of the genus *Fusarium* growing on cereals. This disease has been traced to their trichothecene content. The disease is characterized by damage to the haemopoietic system, which produces the red and white blood corpuscles. In the initial stages there is inflammation and lesions to the skin and, as the disease progresses, the bone marrow atrophies and death follows. A widespread outbreak of the disease in the Southern Urals in Russia in 1944 led to its association with fungal contamination of food cereals. *Fusarium* species will grow at quite low temperatures. In the Ukraine, because of the World War II, the grain was left in the soil over winter and harvested in the spring, by which time it was seriously contaminated, particularly by *F. sporotrichioides*. The metabolites of this fungus caused the death of thousands of people. A disease of cattle known as fescue foot, which is characterized by skin lesions, arises from contact with the snow fungus, *F. nivale*, which grows on grass even in the winter months. This disease is also caused by trichothecenes. Stachybotrytoxicosis is a disease of animals caused by ingestion of food contaminated by the fungus *Stachybotrys atra*. It was responsible for the deaths of many horses in Russia in the 1930s. The symptoms are similar to those of alimentary toxic aleukia. The animals suffer from lesions in the mouth and the alimentary canal. The blood fails to clot and the animal usually dies within a few days. The metabolites that are responsible are known as the satratoxins and are esters of trichothecenes. Fungi are often found growing in damp buildings on materials such as chipboard, wallpaper and ceiling tiles. *Stachybotrys* species such as *S. chartarum* have been identified amongst these organisms. The formation of trichothecenes by *S. chartarum*, which are then ingested *via* airborne dust samples, has been associated with adverse health effects of damp rooms.

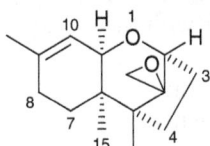

**9.5**

After a wet summer in 1972 corn in parts of the USA was infected with *F. graminearum*. Reports soon followed that pigs refused to eat this corn or had vomited after eating a small amount. The emetic principle, vomitoxin, was identified as 3,7,15-trihydroxy-12,13-epoxytrichothec-9-en-8-one (**9.6**).

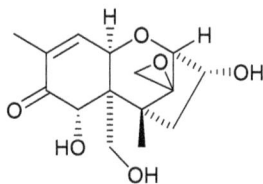

**9.6**

Trichothecin (**9.7**), the first of the trichothecenes to be isolated, was described as an anti-fungal metabolite of the fungus *Trichothecium roseum* in 1949. This fungus produces a pink rot on apples and trichothecin has been associated with the bitter taste of infected fruit. The correct structure of trichothecin embodying its 12,13-epoxide was finally established in 1962 as a result of an inter-relationship with trichodermol. The details of this are described in Chapter 5.

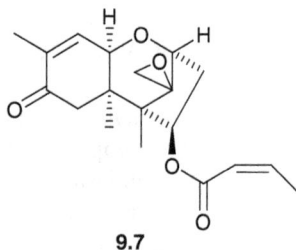

**9.7**

Trichothecenes are characteristic metabolites of *Fusarium* species and have been isolated from species belonging to ten of the twelve sections of the genus *Fusarium* as classified by Booth. The majority contain a 12,13-epoxide and a C-9–C-10 double bond and occur with various combinations of additional oxygen substituents at positions C-3, C-4, C-7, C-8 and C-15. Many of the hydroxyl groups are esterified and there are several macrocyclic esters known as the verrucarins and roridins in which the esters link C-4 and C-15.

The toxic effects of the 12,13-epoxytrichothecenes are associated with the more highly oxidized members such as T-2 toxin (**9.8**) and vomitoxin (deoxynivalenol) (**9.6**) together with their macrocyclic esters. The symptoms of alimentary toxic aleukia in test animals (cats) were detected on administration of T-2 toxin at a level of 0.08 mg kg$^{-1}$ per day. Structure–activity studies on T-2 toxin revealed a requirement for the presence of the 12,13-epoxide and the 9-ene. The compounds exert their activity by the inhibition of protein synthesis. The trichothecenes interfere with the peptidyl transferase in the ribosomes.

The cytotoxic activity of diacetoxyscirpenol (**9.9**) led to its serious consideration as a potential anti-tumour agent but the balance between efficacy and toxicity was too close for a suitable derivative to be obtained.

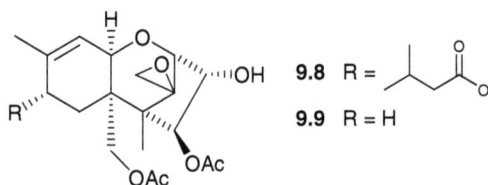

9.8 R =

9.9 R = H

## 9.4 Other *Fusarium* Toxins

*Fusarium* species produce various other toxic metabolites. Fusarin C (**9.10**) is a mutagenic metabolite of *F. monoliforme*. The fumonisins, *e.g.* **9.11**, are another group of mycotoxins that have been isolated from this fungus. They have been associated with a high incidence of esophageal cancer and several animal diseases. The simple molecule monoliformin (**9.12**) is also a toxin.

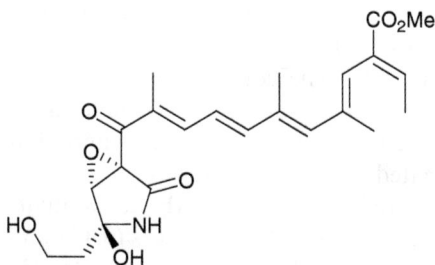

**9.10**

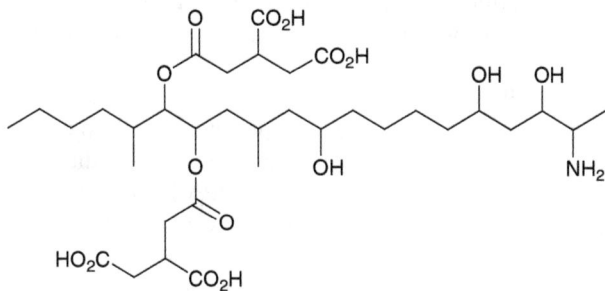

**9.11**

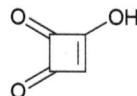

**9.12**

Zearalenone (**9.13**) is a mycotoxin with estrogenic activity and suspected carcinogenicity that is produced by *Fusarium* species such as *F. graminearum* growing on corn. It can cause fertility disorders. The alcohol, zearanol (**9.14**), is used in veterinary medicine as an anabolic agent. In particular it binds to the estrogen receptor and stimulates the synthesis of protein.

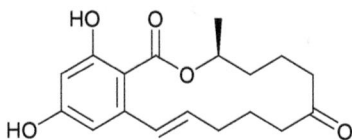

**9.13**

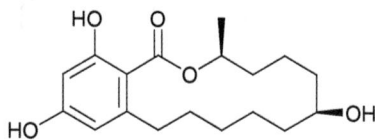

**9.14**

## 9.5  Aflatoxins

Aflatoxins first came to attention as mycotoxins in the 1960s when many thousands of turkeys and ducks, being fattened for the Christmas market, died. The disease was known as 'Turkey X disease'. It first appeared in turkeys in the summer of 1960 and subsequently over 100 000 birds died. The outbreak was characterized by enlarged kidneys and liver lesions. Over 80% of the cases were concentrated within 80–100 miles of London. Investigations showed that the disease was not transmitted between birds but was associated with their feed, all of which had originated from one mill in London. The clue as to the component of the feed that was causing the problem came when a group of cases suddenly appeared in Cheshire. Their feed had come from a mill in Selby. The common constituent with the London mill was a batch of groundnuts from Brazil. Groundnuts were obtained from various countries, including Brazil, Nigeria and India. This particular batch was contaminated with the fungus *Aspergillus flavus*. The aflatoxins were identified as the toxic metabolites. About 20 aflatoxins have now been isolated, of which aflatoxins $B_1$ (**9.15**), $B_2$, $G_1$ (**9.17**) and $G_2$ were the major components. Aflatoxins $B_2$ and $G_2$ are the dihydro derivatives of $B_1$ and $G_1$, respectively. They have subsequently been found in many foods, including edible nuts, oil seeds and cereals on which *A. flavus* has grown as a spoilage organism.

Aflatoxin $B_1$ (**9.15**, Scheme 9.1) is one of the most potent carcinogens known. There is a significant relationship between the presence of aflatoxins in the diet and the incidence of liver cancer. The reactive enol ether of aflatoxin $B_1$ is epoxidized in the liver to form the active metabolite which may react with DNA. The adduct 2-(*N*-7-guanyl)-3-hydroxyaflatoxin $B_1$ (**9.18**) has been detected in the urine of test animals fed aflatoxin. Aflatoxin $B_1$ is also hydroxylated by animals that have ingested it in their fodder. A carcinogenic metabolite, aflatoxin $M_1$ (**9.16**), has been identified in the milk of these animals.

The aflatoxins are biosynthesized by the pathway shown in Scheme 9.1. The intermediate sterigmatocystin (**9.19**) is also carcinogenic.

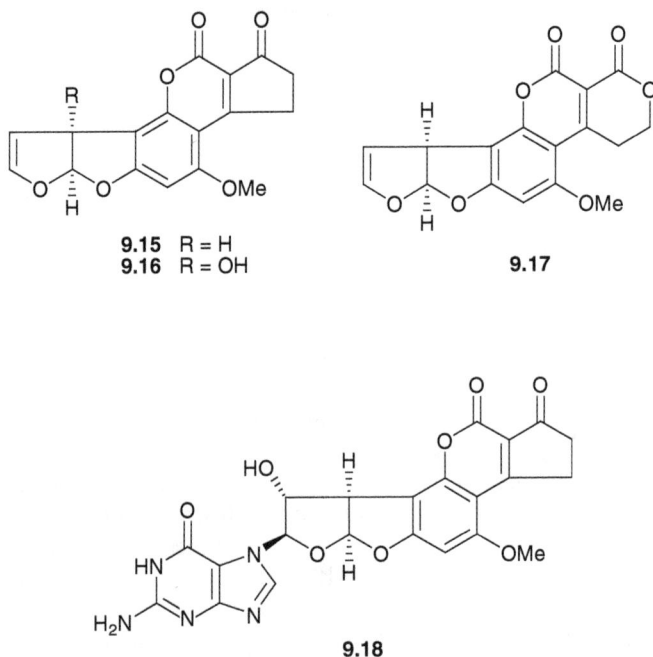

**9.15** R = H
**9.16** R = OH

**9.17**

**9.18**

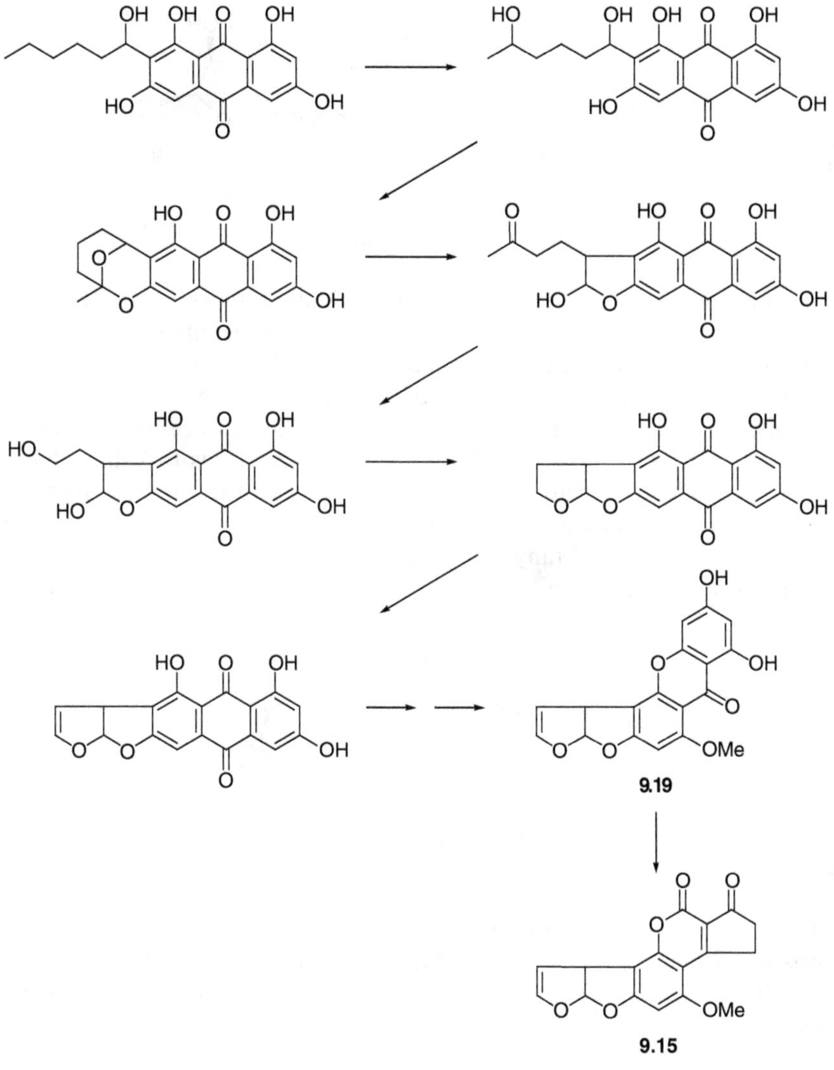

**Scheme 9.1** Biosynthesis of aflatoxins.

## 9.6 Mycotoxins of *Penicillium* Species

*Penicillium* species produce various mycotoxins. Strains of *Penicillium citreo-viride*, which infect rice, produce the mycotoxin, citreoviridin (**9.20**). This compound is the cause of a condition known as 'cardiac beriberi', which is characterized by cardiovascular disorders, difficulty in breathing and nausea. The mechanism of action involves the inhibition of mitochondrial ATP synthesis. Cardiac beriberi was a serious infection of people eating rice as a staple diet, causing many fatalities until the origin was identified.

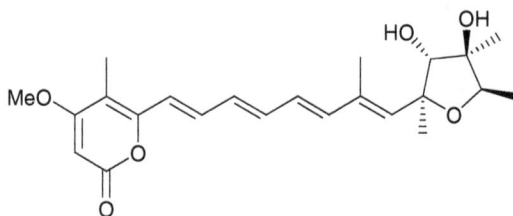

**9.20**

The ochratoxins, *e.g.* **9.21**, which were originally obtained from *Aspergillus ochraceus* in 1965, are produced by various other Aspergillus species and by *Penicillium viridicatum*, which is found on grain. There is evidence that the ochratoxins can occur in a wide range of foods that have been contaminated by these organisms. They have even been detected in pork as a result of carry-over from contaminated animal feed. Ochratoxin A (**9.21**) leads to kidney damage and a high incidence of bladder cancer has been associated with this mycotoxin. It is particularly common in the Balkans, where it is associated with Balkan endemic nephropathy. Many countries have set regulatory limits for the amount of ochratoxin A in food.

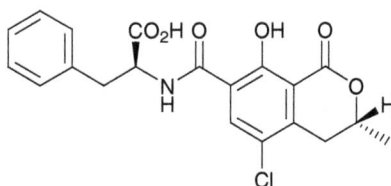

**9.21**

Citrinin (**9.22**) is a widespread mycotoxin that is formed by many *Penicillium* and *Aspergillus* species, particularly *P. citrinum, P. expansum* and *P. roqueforti*. It was first isolated in 1931 by the Raistrick group and described as an antibiotic. It has been shown to have carcinogenic effects and to produce kidney damage. Evidence for its structure and biosynthesis is discussed in Chapter 4.

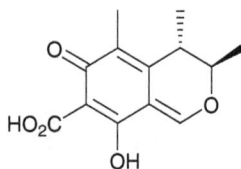

**9.22**

Penicillic acid (**9.23**) was first isolated from cultures of *P. puberulum* and has subsequently been found in many other fungi, particularly from amongst the Penicillia and Aspergilli. It has a range of mutagenic and carcinogenic activities. It is formed by the cleavage of the aromatic ring of 6-methylsalicylic acid (see Chapter 4). Patulin (**9.24**), which is formed by a similar pathway, is a also a metabolite of a range of *Penicillium* species such as *P. patulum, P. expansum* and *P. roqueforti*. It was first isolated from *P. griseofulvum* in the 1940s as an anti-microbial metabolite. However, its toxicity to animals and plants was recognized in the 1960s and it was reclassified as a mycotoxin. It is found in fruit juices. Since *P. expansum* is a spoilage organism of apples, this can be a serious problem and strict limits have been set on its concentration ($50\,\mu\mathrm{g\,kg^{-1}}$). Patulin is highly reactive. It exhibits carcinogenic activity and reacts with various biological nucleophiles, including DNA. It can covalently crosslink proteins by reacting either with the thiol of cysteine or the terminal amino group of lysine. The initial Michael adduct to the $\alpha\beta,\gamma\delta$-unsaturated lactone may undergo ring opening to an aldehyde, which reacts with a further nucleophile.

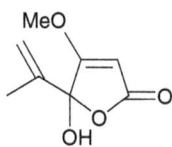 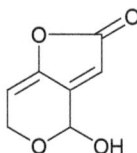

**9.23**            **9.24**

The amino acid tryptophan is a precursor of several fungal metabolites that affect the central nervous system. We have already met it as a constituent of the ergot alkaloids. Roquefortine (**9.25**) is a metabolite of *Penicillium roqueforti* and *P. camemberti*, which are found on some cheeses. Roquefortine is one of a series of mycotoxins that affect the central nervous system and induce tremors. The more complex penitrems are tremorogenic neurotoxins that are produced by the *P. crustosum* series. They are biosynthesized from a tryptophan and a triterpene unit.

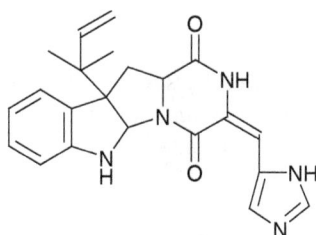

**9.25**

## 9.7 Poisonous Mushrooms

In Europe *Amanita* species, particularly *Amanita phalloides* and to a lesser extent *A. virosa*, are responsible for about 75% of fatal fungal poisonings. Most of these arise as a consequence of mis-identifications.

The toxicity of *Amanita* species is due to the presence of a family of bicyclic octapeptides known as the amanitins (**9.26**). The phallotoxins, which are also present, are heptapeptides. Much of the chemistry of these cyclic peptides was worked out by Wieland. The amanitins have a very high specificity for RNA polymerase II, inhibiting its action and consequently the formation of messenger RNA and protein synthesis. It has been estimated that the $LD_{50}$ in man is 6–7 mg. Unfortunately, the amanitin concentration in an average 60 g *A. phalloides* cap is about 30–60 mg. Amatoxin poisoning follows three stages, an initial period of severe gastrointestinal disturbance about 8–36 hours after ingestion, a period of apparent recovery followed by a terminal phase approximately 3–4 days after ingestion in which liver and kidney failure occur, leading to death.

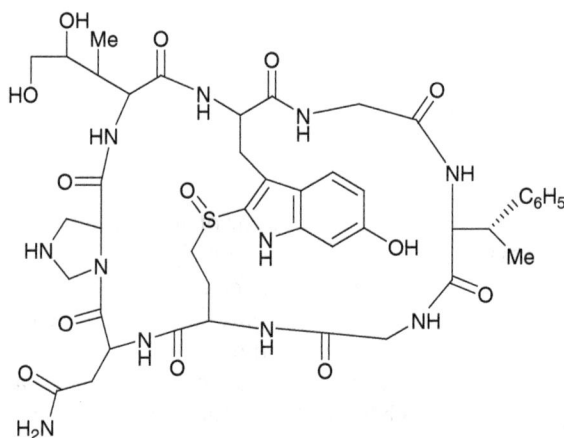

**9.26**

Ingestion of some other members of the genus *Amanita* brings about hallucinations. Two of these are *A. muscaria* (the fly agaric) and *A. pantherina* (the panther mushroom). The use of *A. muscaria* has figured as an inebriant in various religious ceremonies. It has been suggested that *A. muscaria* is the Soma of the Sanskrit hymn known as the Rig Vedas. Their use as intoxicants in the 1700s by people living in Siberia has also been described. The active constituents are a group of isoxazoles, including ibotenic acid (**9.27**) and its decarboxylation product muscimol (**9.28**). These compounds target the γ-aminobutyric acid (GABA) receptors in the brain. GABA is an inhibitory neurotransmitter counterbalancing the effect of various excitatory neurotransmitters. Disturbance of this balance is associated with these isoxazoles. Tricholomic acid (**9.29**), which is produced by *Tricholoma muscarium*, has a similar effect. The action of *A. muscaria* on flies is probably one of inebriation caused by these compounds. However, the fungus also produces muscarine (**9.30**), which is an acetylcholine agonist. This has been used to define one of the sub-sets of the acetylcholine receptors.

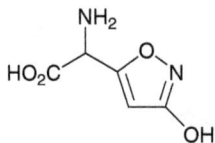

**9.27**

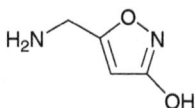

**9.28**

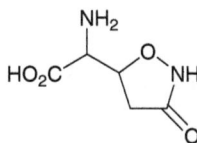

**9.29**

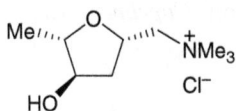

**9.30**

Psilocybin (4-phosphoryloxy-*N*,*N*-dimethyltryptamine) (**9.31**) and the parent alcohol, psilocin, were first isolated from the Mexican mushroom *Psilocybe mexicana* by Hofmann in 1959. Since then these hallucinogenic metabolites have been reported in more than 30 species of this genus and its relatives. The highest concentration (over 0.5% dry weight) of psilocybin has been detected in *Conocybe cyanopus* and *Psilocybe semilanceata*. The hallucinogenic properties of these organisms have led to their colloquial name as 'magic mushrooms'. *P. semilanceata*, which is quite widespread in Europe, is sometimes known as the Liberty Cap. A biosynthetic sequence from tryptophan to psilocybin has been established.

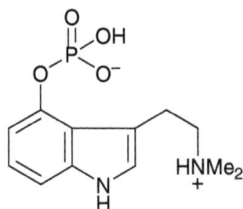

**9.31**

*Amanita* species are not the only toxic mushrooms. Although the false morel *Gyromitra esculenta* is eaten dried or boiled in some countries, the fresh mushroom is poisonous. Raw or incompletely cooked material has caused fatal cases of food poisoning. The major toxin is gyromitrin (**9.32**), which was shown to be acetaldehyde *N*-methyl-*N*-formylhydrazone. It is accompanied by smaller amounts of the *N*-methyl-*N*-formylhydrazones of other aldehydes such as propanal, 3-methylbutanal and hexanal. They are present at a typical level of $57\,\text{mg}\,\text{kg}^{-1}$. Boiling for 10 min reduces this to $1\,\text{mg}\,\text{kg}^{-1}$. Gyromitrin exerts its toxicity by methylating DNA. 7-Methylguanine (**9.33**) has been isolated as a consequence of its action on rats. Gyromitrin decomposes to methylhydrazine, which may undergo oxidation to a methylating agent such as diazomethane.

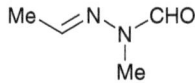

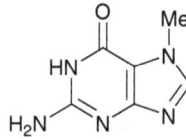

**9.32**                                    **9.33**

When the ink-cap mushroom *Coprinus atramentarius* is eaten alone it is not toxic. However, if it is eaten with alcohol it induces an over-sensitivity that is similar to that of the drug disulphiram (antabuse). The fresh mushroom contains about 160 mg kg$^{-1}$ of the active component, coprine (**9.34**). This was shown to be $N^5$-(1-hydroxycyclopropyl)-L-glutamine, which contains the unusual *N*-acyl-1-aminocyclopropanol unit. It is metabolized to L-glutamic acid and cyclopropanone. The hydrate of cyclopropanone is a good inhibitor of acetaldehyde dehydrogenase. This induces elevated levels of acetaldehyde in the blood and retards the rate of ethanol metabolism.

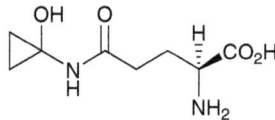

**9.34**

# CHAPTER 10
# *Fungi as Reagents*

## 10.1 Introduction

The biotransformation of organic compounds by microorganisms has been an important aspect of microbiological chemistry throughout its development. In 1858 Pasteur used the fungus *Penicillium glaucum* to obtain L-ammonium tartrate from DL-ammonium tartrate by the selective destruction of the D-enantiomer. Today microbiological transformations are used in the commercial production of several industrial chemicals.

It is possible to distinguish between two major types of biotransformation. On the one hand, there are xenobiotic transformations in which the substrate is completely alien to the microorganism whilst, on the other hand, there are biosynthetically-patterned biotransformations in which the substrate bears a formal relationship to an intermediate of one of the natural biosynthetic pathways that the organism possesses.

## 10.2 Xenobiotic Transformations

Biotransformations have several valuable synthetic attributes when compared to chemical processes. They are often regio-, stereo- and enantiospecific and they may not require the use of protecting groups. Their regiospecificity can complement that of chemical reactions and in some instances they provide access to sites in a molecule that are chemically difficult to reach. Biotransformations can be carried out under mild, environmentally friendly conditions whilst the reagent, the microorganism, is self-replicating. However, there are disadvantages. Large volumes of medium can be involved and yields, particularly in the initial stages of an investigation, may be disappointingly poor. Various techniques using immobilized organisms and organic co-solvents have been developed to overcome some of these problems. Although several rules

The Chemistry of Fungi
By James R. Hanson
© James R. Hanson, 2008

and models exist to predict the structures of biotransformation products, the outcome of a biotransformation is often difficult to predict with certainty.

Biotransformations may be carried out with isolated enzyme systems or with intact organisms. Although isolated enzyme systems may be more specific and efficient for certain biotransformations, these reactions may involve isolating the enzyme system and, for some classes of enzyme-catalysed reaction, a recycling sequence may be required to regenerate the enzyme. In these cases whole organisms may be more appropriate. Although many biotransformations of xenobiotic substances have been observed, the common reactions are hydrolysis, redox reactions, hydroxylations, epoxidations and aldol condensations. Various bacteria, yeast and fungi and enzyme systems derived from mammalian sources such as pig's liver have been employed for different biotransformations. Although this book is concerned with fungi, it is important to remember that these are by no means the only systems that have been used for biotransformations.

## 10.2.1 Microbial Hydrolysis

An important property of enzyme reactions is their chiral nature and this is reflected in the use of microorganisms in the hydrolysis of esters to obtain a chiral product. One of the consequences of the thalidomide tragedy of the 1960s is that a racemic drug has to be resolved and the biological activity and potential toxicity of the individual enantiomers has to be assessed. The introduction of chirality and the resolution of racemic mixtures are important aspects of the synthesis of chiral drugs. Various proteases, lipases and esterases are commercially available for use in the hydrolysis of a range of substrates. Systems that are of fungal origin include preparations from *Rhizopus* species, *Gliocladium roseum* and the yeast *Candida cylindracea*. Many of these hydrolyses are kinetic resolutions rather than complete resolutions. Consequently, some experimentation may be required to obtain the highest enantiomeric excess. Although a hydrolysis may be used just to resolve a racemate, it may also be used to introduce chirality by the enantiospecific distinction between two paired chemically identical ester groups attached to a pro-chiral centre in a non-chiral starting material. An example (Scheme 10.1) is provided by the enantiospecific hydrolysis of dimethyl-3-hydroxy-3-methylglutarate (**10.1 → 10.2**) in a synthesis of chiral mevalonolactone (**10.3**).

The results of the hydrolysis of the esters of alcohols by several lipases, including those from *Candida* species, have been rationalized in terms of a

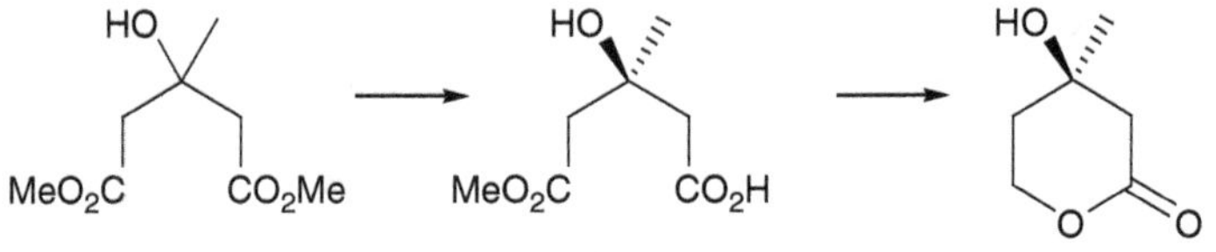

**Scheme 10.1**  A stereospecific synthesis of mevalonolactone.

**Scheme 10.2**  Stereochemistry of hydrolyses by *Candida rugosa* esterase.

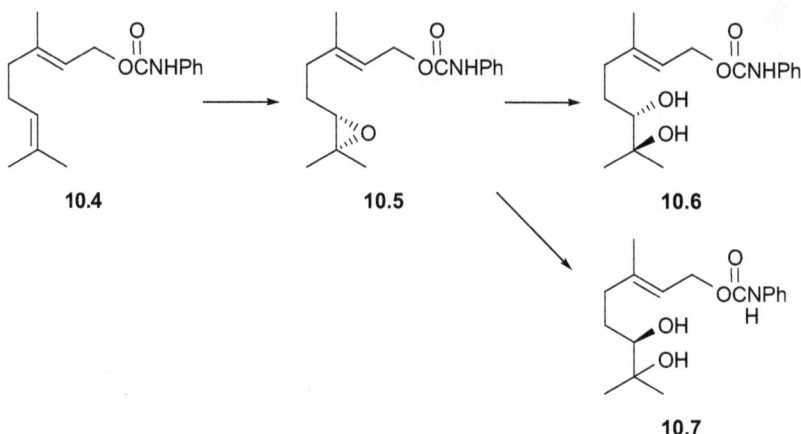

| 10.4 | 10.5 | 10.6 |

10.7

**Scheme 10.3**  Microbial formation and hydrolysis of an epoxide.

predictive rule. The stereochemistry of the enantiomer that will be formed preferentially on hydrolysis of an ester is shown in Scheme 10.2, in which M = a medium sized substituent and L = a large substituent.

The microbial hydrolysis of epoxides can also be a useful synthetic method for introducing chirality. The example shown in Scheme 10.3 involved epoxidation of the 6,7-double bond of geraniol N-phenylcarbamate (**10.4**) by the fungus, *Aspergillus niger* to give the (6S)-epoxide (**10.5**). This epoxide underwent a spontaneous acid-catalysed hydrolysis at pH 2 to form the (6S)-diol (**10.6**). However, at pH 5–6 this underwent enzymatic hydrolysis to give the (6R)-diol (**10.7**).

## 10.2.2  Microbial Redox Reactions

The reduction of a ketone to create a chiral secondary alcohol is one of the most widely used biotransformations. These enzymatic reductions take place under mild conditions with an enantiomeric homogeneity that can be difficult to achieve by chemical means. Both intact microorganism and isolated enzyme systems have been used for these reductions, although the latter, which may utilize the nicotinamide co-enzymes, will require a further recycling stage. Bakers' yeast is widely used as an intact microorganism, although many fungi, *e.g. Curvularia falcata*, have also been employed. The stereochemical outcome of many of the reductions follows a pattern that has been summarized in Prelog's rule. This rule, which takes into account the steric effects of the substituents adjacent to the carbonyl group, is illustrated in the Scheme 10.4.

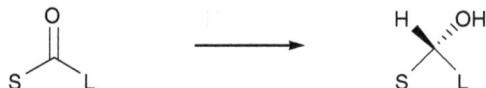

**Scheme 10.4**  Prelog's rule to predict the stereochemistry of the microbial reduction of a ketone.

Where there is a chiral centre adjacent to the ketone it may also influence the stereochemistry of the reduction. Thus the reduction of ethyl acetoacetate (**10.8**) by yeast has the stereochemical result shown in **10.9**.

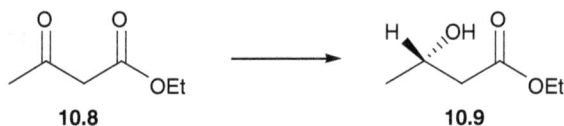

## 10.2.3   Microbiological Hydroxylation

A major step forward in the use of microorganisms for biotransformations came in their application to the 11-hydroxylation of steroids. The cortical steroids, exemplified by cortisone, have an oxygen function at C-11. They are not readily available from animal sources in sufficient quantity for medicinal uses. The biological activity of these steroids in alleviating the pain and inflammation associated with rheumatoid arthritis puts a high premium on their successful partial synthesis from more readily available plant sterols. Whereas a steroid hormone such as progesterone can be synthesized chemically from plant sterols such as diosgenin, the insertion of an oxygen function at C-11 was chemically difficult. Even the transposition of an oxygen function from C-12 to C-11 starting from hecogenin or the bile acids involved several steps. In 1952 it was found that the incubation of progesterone (**10.10**) with the fungus *Rhizopus arrhizus* led to the 11α-hydroxylation (**10.11**) in excellent yield. Subsequently, many other steroids have been incubated with this and other fungi. The site of hydroxylation depends on the site of existing substituents and the microorganism that has been used. Almost all centres on the steroid framework have been hydroxylated. Common sites for the hydroxylation of progesterone and its relatives are C-6β and C-11α. A very thorough investigation by Jones of the hydroxylation of over 300 androstanes using several organisms, including *Calonectria decora*, *Rhizopus nigricans* and *Aspergillus ochraceus*, was carried out during the late 1960s and 1970s. It was found that the stereochemistry of the A/B ring junction was a very important factor in influencing the interaction of the steroid substrate with the steroid hydroxylase. The 5α-androstanes were the most intensively studied. Compounds bearing carbonyl or hydroxyl substituents at various sites were investigated. The commonest sites to be hydroxylated were C-6β, C-7α, C-7β, C-11α, C-12α, C-12β, C-15α and C-16β. Surprisingly, bearing in mind the androstane–estrogen relationship, microbial hydroxylation at C-19 is rare.

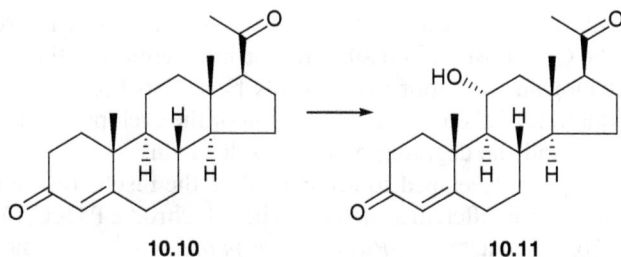

**10.10**   **10.11**

A model to accommodate the hydroxylation of 5α-androstane by *Calonectria decora* was proposed by Jones. The model envisaged a triangular relationship between two binding sites on the steroid substrate such as hydroxyl or carboxyl groups and the site of hydroxylation. The average distances are set out in **10.12**. However, the two-point binding of the relatively flat 5α-androstane could occur in as many as four different orientations, giving rise to several different products. These orientations are described as normal, reverse, normal capsized or reverse capsized (Scheme 10.5). The capsized orientations were obtained by rotation of the steroid molecule about the C(3)–C(17) axis by 180°. These different orientations were not equally favoured. For example, many of the anticipated products from the capsized orientations were not detected. Structural variations such as increasing the size of ring D to a six-membered ring favoured particular orientations, in this case the reverse binding.

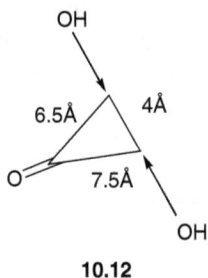

**10.12**

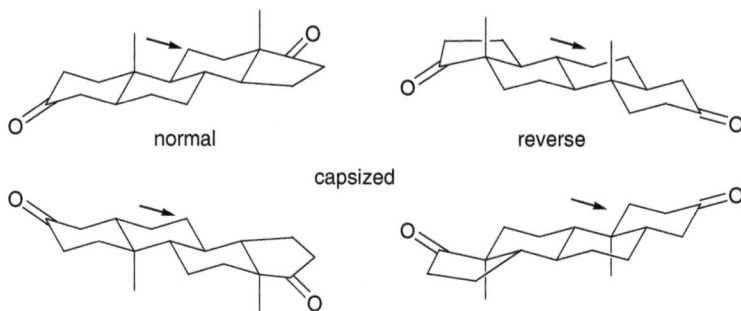

normal    reverse

capsized

**Scheme 10.5** The different binding modes of a steroid in a microbiological hydroxylation.

Steroidal unsaturated ketones such as the 3-keto-4-enes were hydroxylated in the axial allylic C-6β position, possibly *via* the intervention of the enol. Several other fungal biotransformations of steroids have also been observed. These include epoxidation of alkenes, the conversion of the cyclopentanone of ring D into a δ-lactone and the degradation of the side-chain.

Models have been developed to accommodate the results of the hydroxylation of substrates with different structures. The cytochrome P450CAM camphor hydroxylase from the bacterium *Pseudomonas putida* has been studied by X-ray crystallography. The importance of hydrophilic interactions with a valine (VAL-247) and a polar interaction mediated by hydrogen bonding to a tyrosine residue (TYR-96) has been noted. A model based on the hydroxylation of numerous cyclic amides by *Beauveria sulfurescens* (originally named *Sporotrichum sulfurescens*) showed that hydroxylation occurred preferentially at a methylene group about 5.5 Å from an electron-rich substituent on the substrate.

The relatively flat 5α-androstanes did not give much information on the three-dimensional topology of a hydroxylase. Models have been devised based on the results of hydroxylation of polycyclic sesquiterpenes that contain bridged ring systems. One such model (**10.13**) has been devised for the fungus *Mucor plumbeus*. These models were created by superimposing structures using an axis between the putative binding site of the molecule and the site of hydroxylation together with a calculated 'centre of gravity' of the molecule. A diverse series of structures were superimposed and a set of 1-Å cubic boxes built around the conglomerate structure. Boxes that did not contain any atoms were deleted to reveal the boundaries of an overall three-dimensional model of the hydroxylase. The model successfully predicted the sites of hydroxylation of some sesquiterpenoids that were not included in the construction of the original model.

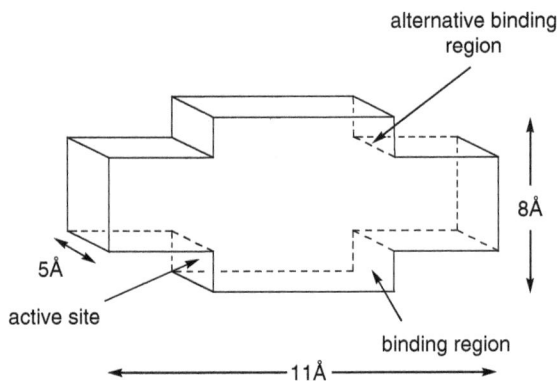

**10.13**

Fungi will hydroxylate an aromatic ring *via* the formation of an arene oxide. The biotransformation of aromatic substrates by the fungus *Aspergillus niger* has been examined. Typical of aromatic hydroxylations by mono-oxygenases, the reactions are accompanied by a rearrangement of hydrogen, the NIH shift. Thus, 4-deuterioanisole (**10.14**) gave 3-deuterio-4-hydroxyanisole (**10.15**).

The metabolism of the herbicide 2,4-dichlorophenoxyacetic acid by *A. niger* has been examined. The major metabolite was 2,4-dichloro-5-hydroxyphenoxy-acetic acid. This was accompanied by a minor product, 2,5-dichloro-4-hydroxy-phenoxyacetic acid, in which a chlorine migration had taken place.

10.14          10.15

# 10.3 Biosynthetically-patterned Biotransformations

Biosynthetically-patterned microbiological transformations exploit the substrate flexibility of enzymes involved in the biosynthesis of secondary metabolites. These biotransformations are sometimes known as analogue biosynthesis or precursor-directed biosynthesis. This approach to biotransformation can be useful in preparing analogues of biologically-active microbial metabolites for structure–activity studies, a feature that has been exploited with penicillins using a cloned isopenicillin N synthase. The structures of the substances that are transformed and of their products can also shed light on the stereo-electronic constraints of enzymatic steps and on the nature of biosynthetic intermediates, a feature that has been exploited in studies on the cyclization of squalene to the triterpenes and steroids.

However, there are some problems that are inherent in using this approach to biotransformation. Firstly, a xenobiotic transformation often involves just one step, *e.g.* a hydroxylation, whilst a biosynthetically-directed transformation may use a sequence of enzymatic steps. The artificial substrate may be a poor fit for some of these specific enzymes. Consequently, the overall yields may be rather poor. Secondly, unless the fermentation is modified in some way, the artificial exogenous substrate may be in competition with the natural metabolites, di-minishing the yield and posing separation problems. Nevertheless, in several cases biosynthetically-directed biotransformations have been exploited to yield novel biologically-active compounds and useful biosynthetic information.

Several biosyntheses involve assembling building blocks such as amino acids and carboxylic acids. Parts of these assembly processes can show some sub-strate flexibility, allowing the enzymatic synthesis of un-natural analogues of normal fungal metabolites. One of the earliest examples was the variation of the penicillin side-chain brought about by replacing the phenylacetic acid of penicillin G with phenoxyacetic acid as in penicillin V.

In several instances the amino acids of biologically active fungal peptides have been replaced by analogues. The cyclosporins are a group of closely rela-ted cyclic undecapeptides that are produced by the fungus *Tolypocladium inflatum* (*Beauveria inflata*). These metabolites have attracted interest as immunosuppressive agents. Supplementation of the fermentation medium with amino acids such as L-threonine or L-valine modified the relative amounts of different members of the cyclosporin complex that were formed by the fungus. Administration of analogues of the natural amino acids such as DL-α-allylglycine afforded novel cyclosporin analogues.

The ergot alkaloids produced by *Claviceps purpurea* consist of a lysergic acid and a peptide moiety. Feeding L-norvaline to a strain of *C. purpurea* gave rise to modified alkaloids such as ergorine and ergonorine. Modification of the tryptophan precursors of the asperlicin benzodiazepine metabolites of *Aspergillus alliaceus* resulted in the formation of asperlicin analogues. The beauvericins and enniatins are cyclohexadepsipeptides that are produced by *Paecilomyces tenuipes* or *Verticillium hemipterigenum* and which possess insecticidal and phytotoxic activity. By varying the amino acid composition of the medium on which these fungi were grown by the addition of, for example, L-isoleucine it was possible to change the composition of the beauvericins and enniatins that were produced.

Although it is difficult to vary the internal components of the polyketide chain assembled by polyketide synthase, a range of successful variations have been produced by modification of the starter unit, particularly with *Streptomycete* metabolites. A propionyl co-enzyme A starter unit is used by *Saccharopolyspora erythraea* in the biosynthesis of erythromycin. Replacement of this by 3-fluoropropionate or 4-fluorobutyrate analogues gave rise to fluoro-erythromycin. The macrolide soraphen A is an antifungal metabolite of *Streptomyces cellulosum*. Non-natural benzoyl CoA starter units derived from fluorinated phenylalanine, cinnamate and benzoates gave rise to a series of soraphen analogues.

Biotransformations with the fungus *Gibberella fujikuroi* illustrate many uses of this methodology. Gibberellic acid, the major metabolite of this fungus, is the best known of the gibberellin plant hormones. About 140 gibberellin plant hormones are now known. They share a common carbon skeleton and differ from one another in the substituents on their framework. Whilst gibberellic acid is produced in a relatively large amount by the fungus, most of the gibberellins have only been isolated in very small amounts from higher plants.

Details of the biosynthesis of gibberellic acid by *Gibberella fujikuroi* are discussed in Chapter 5. In broad outline the pathway (Scheme 10.6) involves the oxidation of the hydrocarbon ent-kaurene (**10.16**) at C-19, hydroxylation at C-7β to form ent-7α-hydroxykaurenoic acid (**10.17**), and ring contraction to gibberellin $A_{12}$ 7-aldehyde (**10.18**). The latter is converted *via* several gibberellins into gibberellic acid (**10.19**) by hydroxylation at C-3β (**10.20**), oxidative removal of C-20 and the formation of the γ-lactone ring, the introduction of the ring A double bond and hydroxylation at C-13. The diterpenoid biosynthetic pathways in *G. fujikuroi* diverge at various points, leading to the kaurenolide lactones (**10.22**), the aldehyde–anhydride fujenal (**10.23**) and to some other gibberellins, *e.g.* **10.21**.

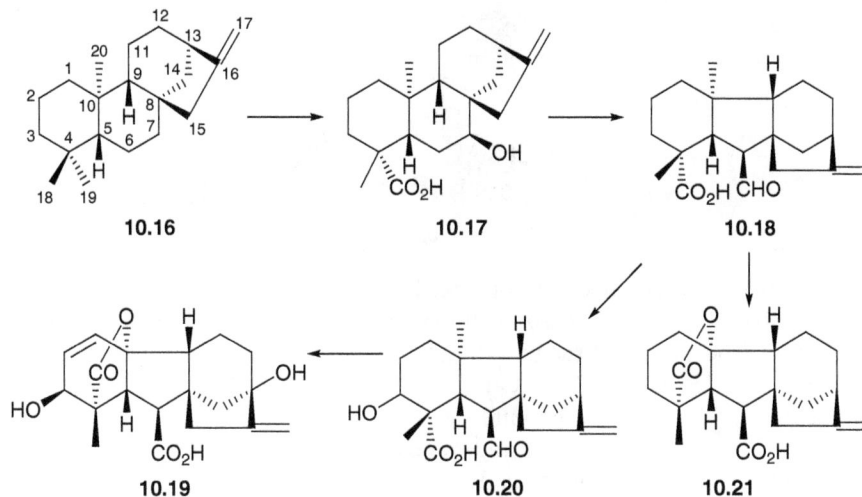

**Scheme 10.6** Biosynthesis of gibberellic acid.

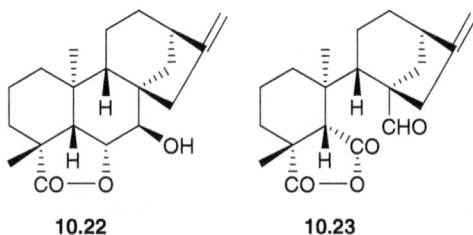

A major problem in these biosynthetically-directed transformations is the separation of the normal and artificial metabolites. Two strategies have been used to overcome these problems. The first employed a mutant Bl-41a, which was blocked at the stage of the oxidation of ent-kaurene at C-19, an early stage in gibberellin biosynthesis. In the second method, the fungus was grown in the presence of compounds such as chlorocholine chloride (CCC) and AMO-1618, which inhibit kaurene synthase. Both methods ensured that the fungus did not produce substantial quantities of endogenous post-kaurene metabolites.

In the fungus, hydroxylation at C-13 is normally the last stage in the biosynthesis of gibberellic acid. However, in some higher plants hydroxylation at C-13 takes place at a much earlier stage in gibberellin biosynthesis and, consequently, there are several rare plant gibberellins related to the fungal gibberellin intermediates but also possessing a C-13 hydroxyl group. Steviol, ent-13-hydroxykaurenoic acid (**10.24**), is a relatively readily available ent-kaurenoic acid analogue. Its biotransformation by *G. fujikuroi* made available a series of 13-hydroxylated plant gibberellins.

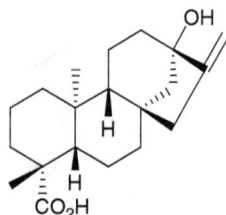

**10.24**

Variously hydroxylated ent-kaurenes have been obtained from higher plants and incubated with *G. fujikuroi* with the object of making rare gibberellins and of defining the 'width' of the biosynthetic pathway. Compounds with hydroxyl groups at C-2α, C-2β, C-3β, C-14β, and C-15β were transformed into $C_{19}$ gibberellins. However, compounds with a C-11 oxygen function were only converted into $C_{20}$ gibberellins and the oxidative removal of C-20 did not appear to take place. The presence of a C-3α hydroxyl group blocked hydroxylation at C-19 whilst a C-18 hydroxyl group appeared to prevent oxidation at C-6β and ring contraction. The presence of a hydroxyl group in the substrate that was adjacent to a site of metabolism introduced some constraints in the biotransformation. The hydration of the 16-ene to give a C-16 tertiary alcohol acted as a 'dumping' mechanism, preventing further transformation of these compounds.

The inhibitory effect of an extra hydroxyl group adjacent to a site of metabolism has been used to provide evidence for a putative biosynthetic intermediate. The kaurenolide lactones are formed from ent-kaur-6,16-dien-19-oic acid (**10.25**) by *G. fujikuroi*. It was suggested that a 6,7-epoxide was formed first and that this underwent a diaxial opening with participation of the 19-carboxylic acid to generate the lactone ring. However, it was not possible to detect the formation of the epoxide, presumably due to the rapid hydrolysis. Incubation of ent-kaur-6,16-diene with *G. fujikuroi* gave 7-hydroxykaurenolide whilst ent-18-hydroxykaur-6,16-diene gave 7,18-dihydroxykaurenolide. Use was made of the inhibiting effect of a C-3α hydroxyl group on the oxidation of C-19. Incubation of the corresponding C-3α hydroxylated kaur-6,16-diene (**10.26**) gave the corresponding 6β,7β-epoxides without oxidation having taken place at C-19.

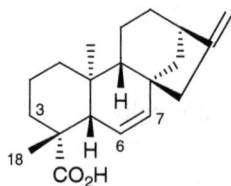

**10.25**

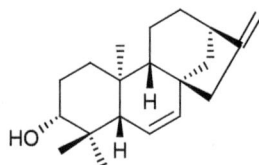

**10.26**

The later stages of a biosynthetic pathway often introduce a chemical complexity that can make structural modifications of the final metabolite difficult. The introduction of methyl substituents onto 13-hydroxylated gibberellins was achieved by feeding 1- and 2-methylgibberellin A4 (**10.27**) to *G. fujikuroi* and allowing the fungus to introduce the 13-hydroxyl group. [17-$^{14}$C]Gibberellic acid was obtained by feeding the much more easily prepared [17-$^{14}$C]gibberellin derivative (**10.28**).

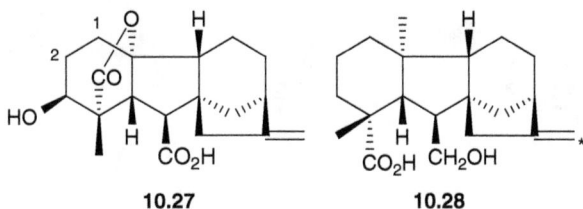

**10.27**                    **10.28**

The carbon skeleton of the ent-kaurenoid precursors of the gibberellins is closely related to that of the diterpenoids of the ent-atiserene (**10.29**), ent-beyerene (**10.30**) and ent-trachylobane (**10.31**) series. The differences lie in the structure of ring D. The flexibility of the gibberellin biosynthetic pathway in *G. fujikuroi* has been explored with substrates possessing these different carbon skeleta, affording atisa-, beyer- and trachyloba-gibberellins.

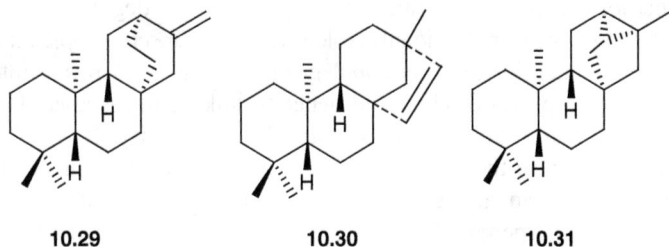

**10.29**              **10.30**              **10.31**

Whilst these biosynthetically-patterned biotransformations of terpenoids have been described in the context of studies with *G. fujikuroi*, studies with other fungi have shown that a similar pattern of results can be obtained using other biosyntheses. Although the yields of the metabolites are often poor, the transformations do have some value in the preparation of rare or labelled compounds. Furthermore, the results reveal some constraints on the biosynthetic pathways and provide information on the stereo-electronic requirements of enzymatic processes that could not be obtained from conventional studies.

# *Epilogue*

The aim of this book has been to describe some aspects of the diverse chemistry of fungal metabolites, their biosynthesis and their biological activity. The way in which the elucidation of the structure of fungal metabolites has developed from the 1930s to the present day reveals the immense impact of spectroscopic methods on organic chemistry. There are major differences between the strategies that were used up to the 1950s and those that are used at the present time. The earlier chemical methods relied on structural simplification and a degradation that broke the molecule down into smaller identifiable fragments. Spectroscopic methods are used not only to identify the functional groups but also to reveal interactions between parts of a natural product and hence to link them together. These strategies rely on structural complexity to establish connectivity. It is instructive in this context to compare the work that was carried out on the elucidation of the structure of trichothecin and gibberellic acid with more recent studies on, for example, the sesquiterpenes of the Basidiomycetes. Another instructive comparison is between penicillin, cephalosporin C and the Streptomycete product clavulanic acid. However, it is also worth reflecting on the novel heterocyclic chemistry that was discovered during the studies on the structure of the penicillins and the rearrangements that were uncovered during the work on the trichothecenes and the gibberellins.

Fungi have several advantages over plants for biosynthetic studies in terms of the incorporation of precursors and the relative ease of isolation of biosynthetic enzymes. The production of fungal metabolites is not constrained by the seasons. Biosynthetic studies have benefited from the development of spectroscopic methods and new experiments have become possible. In polyketide biosynthesis it is instructive to compare the laborious chemical degradations that were required to establish the specificity of labelling in studies that were undertaken in the 1950s and early 1960s on 6-methylsalicylic acid and griseofulvin with studies using stable isotopes and NMR methods of detection that were reported in the 1970s and 1980s with metabolites such as terrein and asperlactone.

Whilst many fungal metabolites such as penicillin, griseofulvin and trichothecin were originally discovered because of their anti-bacterial or anti-fungal properties, the widening scope of bio-assays has led not only to the discovery of new metabolites such as the statins but also to a re-evaluation of known compounds such as mycophenolic acid and pleuromutilin against other bio-assays. The cyclosporins, now used as immunosuppressive agents, were originally discovered as antibiotics. There is a case for re-examining many of the metabolites that were isolated 50 or more years ago using modern screening methods. The identification of the β-methoxyacrylate anti-fungal pharmacophore in the strobilurin metabolites of the Basidiomycetes led to the development of novel synthetic anti-fungal agents containing this moiety.

There is an increasing understanding of the ecological role that fungal metabolites play in establishing the niche that particular microorganisms occupy in nature. These compounds may defend the organism against attack or facilitate the dissemination of its spores. They may act as phytotoxins or as mycotoxins. Efforts to understand the ecological chemistry of the sesquiterpenoid metabolites of the Basidiomycetes, the spread of Dutch elm disease or the role of the metabolites of a plant pathogen such as *Botrytis cinerea* have provided the stimulus for chemical studies. The gibberellin plant hormones were originally discovered as the metabolites of a rice pathogen, *Gibberella fujikuroi*. There have been many developments in the microbiological chemistry involved in the isolation, detection and mechanism of action of mycotoxins such as the aflatoxins, the trichothecenes, patulin, the tremorogenic inindole alkaloids of *Penicillium* species and the ergot alkaloids. The disease implications of the presence of some fungal metabolites arising from the spoilage of food are an on-going area of analytical and microbiological chemistry.

The biosynthetic and biodegradative abilities of fungi have led to their application as environmentally acceptable reagents. The stereo- and regiospecific nature of enzymatic catalysis can complement that of chemical reagents and widen the armoury of the chemist. The substrate flexibility of the later stages of biosynthesis such as that of the penicillins, the cyclosporins and the gibberellins opens the way for the synthesis of 'un-natural' natural products of pharmaceutical and phytochemical interest. Molecular modelling to establish the 'width' of parts of a biosynthetic pathway and to understand the role of 'dumping' mechanisms, such as the hydration of the 16-ene in the kaurenoid metabolism of *Gibberella fujikuroi*, may provide a clue as to why some seemingly bio-inactive fungal metabolites occur.

The understanding that is now emerging of the genetics of the biosynthetic pathways leading to secondary metabolites provides many opportunities for the microbiological chemist to become involved in studies of the regulation of fungal metabolite biosynthesis. The fact that many of these genes are clustered and that their expression may be determined by transcriptional regulatory factors has many implications for microbiological chemistry. The future for new chemical forays into the complex world of the fungi is as bright as ever.

# Further Reading and Bibliography

There are about 200 000 known natural products, of which probably a quarter are fungal metabolites. Fortunately, there are databases such as the *Dictionary of Natural Products* and *Natural Product Updates* which list these (for a review of these databases see, M. Fullbeck, E. Michalsky, M. Dunkel and R. Preissner, *Nat. Prod. Rep.*, 2006, **23**, 347). Hence this bibliography contains just the titles of some books and reviews together with a few papers of historical interest or which afford an entry into the literature concerning particular fungal metabolites. In addition to those articles that are listed under particular chapters in the bibliography, there are other series of regular articles in *Natural Product Reports* on families of natural products which include fungal metabolites. Each of these gives the references to the previous reports, examples are: The biosynthesis of plant alkaloids and nitrogenous microbial metabolites, R.B. Herbert, *Nat. Prod. Rep.*, 2003, **20**, 494; Simple indole alkaloids and those with a non-rearranged monoterpenoid unit, T. Kawasaki and K. Higuchi, *Nat. Prod. Rep.*, 2005, **22**, 761; Marine derived fungi – a chemically diverse group of micro-organisms, T.S. Bugni and C.M. Ireland, *Nat. Prod. Rep.*, 2004, **21**, 143; The biosynthesis of marine natural products: microorganisms, B.S. Moore, *Nat. Prod. Rep.*, 2005, **22**, 580. Natural sesquiterpenoids, B.M. Fraga, *Nat. Prod. Rep.*, 2006, **23**, 943; Diterpenoids, J.R. Hanson, *Nat. Prod. Rep.*, 2006, **23**, 875; Triterpenoids, J.D. Connolly and R.A. Hill, *Nat. Prod. Rep.*, 2007, **24**, 465.

## Books

*Fungal Metabolites*, W.B. Turner, Academic Press, London, 1971; *Fungal metabolites II*, W.B. Turner and D.C. Aldridge, Academic Press, London, 1983.

*ROMPP Encyclopedia of Natural Products*, ed. W. Steglich, B. Fugmann and S. Lang-Fugmann, Thieme, Stuggart, 2000.

*Chemical Microbiology*, A.H. Rose, Butterworths, 2nd edn., London, 1968.

*Smith's Introduction to Industrial Mycology*, A.H.S. Onions, D. Allsopp and H.O.W. Eggins, Edward Arnold, 7th edn., London, 1981.

*Ainsworth and Bisby's Dictionary of the Fungi*, ed. G. Ainsworth, 6th edn., Commonwealth Mycological Institute, Kew, 1971.

*Mushrooms and other Fungi of Great Britain and Europe*, R. Phillips, Pan Books, London, 1981.

*Biology of Micro-organisms*, L.F. Hawker, A.H. Linton, B.F. Folkes and M.J. Carlile, Edward Arnold, London, 1960.

## Chapter 1

Phenotypic taxonomy and metabolite profiling in microbial drug discovery, T.O. Larsen, J. Smedsgaard, K.F. Nielsen, M.E. Hansen and J.E. Frievad, *Nat. Prod. Rep.*, 2005, **22**, 672.

Twenty five years of chemical ecology, J.B. Harborne, *Nat. Prod. Rep.*, 2001, **18**, 361.

Biogenetic speculation and biosynthetic advances, R. Thomas, *Nat. Prod., Rep.*, 2004, **21**, 224.

From iso, sake and shoyu to cosmetics; a century of science for kojic acid, R. Bentley, *Nat. Prod. Rep.*, 2006, **23**, 1046.

## Chapter 2

Studies in the biochemistry of micro-organisms, H. Raistrick, *Philos. Trans. R. Soc. (B)*, 1931, **220**, 1.

Metabolism of *Gibberella fujikuroi* in stirred culture, A. Borrow, E.G. Jefferys, R.H.J. Kessell, E.C. Lloyd, P.B. Lloyd and I.S. Nixon, *Can. J. Microbiol.*, 1961, **7**, 227.

Studies in the biochemistry of micro-organisms, 65, A survey of chlorine metabolism by moulds, P.W. Clutterbuck, S.L. Mukhopadhyay, A.E. Oxford and H. Raistrick, *Biochem. J.*, 1940, **34**, 664.

Studies in the biochemistry of micro-organisms 77, A survey of fungal metabolism of inorganic sulfates, H. Raistrick and J.M. Vincent, *Biochem. J.*, 1948, **43**, 90.

Chloromethane biosynthesis in poroid fungi, D.B Harper, J.T. Kennedy and J.T.G. Kamilton, *Phytochemistry*, 1988, **27**, 3147.

Formation of organometalloid compounds by micro-organisms, trimethylarsine and dimethylethylarsine, F. Challenger, C. Higginbottom and L. Ellis, *J. Chem. Soc.*, 1933, 95.

A review of trace element concentrations in edible mushrooms, P. Kalac and L. Svobada, *Food Chem.*, 2000, **69**, 273.

Methodology (in biosynthetic studies), S.A. Brown, *Biosynthesis*, Specialist Periodical Reports, Royal Society of Chemistry, London, 1972, vol. 1, p. 1.

Some new NMR methods for tracing the fate of hydrogen in biosynthesis, M.J. Garson and J. Staunton, *Chem. Soc. Rev.*, 1979, **8**, 539.

Applications of multinuclear NMR to structural and biosynthetic studies of polyketide microbial metabolites, T.J. Simpson, *Chem. Soc. Rev.*, 1987, **16**, 123.

Natural products of filamentous fungi: enzymes, genes and their regulation, D. Hoffmeister and N.P. Keller, *Nat. Prod. Rep.*, 2007, **24**, 393.

# Chapter 3

## Section 3.2

Sir Robert Robinson and the early history of penicillin, E.P. Abraham, *Nat. Prod. Rep.*, 1987, **4**, 41.

The chemistry of the penicillins, A.H. Cook, *Quart. Rev.* 1948, **2**, 203.

The formation from glucose by members of the *Penicillium chrysogenum* series of pigments, an alkali soluble protein and penicillin, the anti-bacterial substance of Fleming, P.W. Clutterbuck, R. Lovell and H. Raistrick, *Biochem. J.*, 1932, **26**, 1907.

## Section 3.3

The structure of Cephalosporin C, E.P. Abraham and G.G.F. Newton, *Biochem. J.*, 1961, **79**, 377.

## Section 3.4

The biosynthesis of penicillins and cephalosporins, J.E. Baldwin and E. Abraham, *Nat. Prod. Rep.*, 1988, **5**, 129.

The β-lactamase cycle: a tale of selective pressure and bacterial ingenuity, A. Matagne, A. Dubus, M. Galleni and J.M. Frere, *Nat. Prod. Rep.*, 1999, **16**, 1.

New penicillins from isopenicillin N synthase, J.E. Baldwin, M. Bradley, S.D. Abbott and R.M. Adlington, *Tetrahedron*, 1991, **47**, 5304.

Isopenicillin N synthase, mechanistic studies, J.E. Baldwin and M. Bradley, *Chem. Rev.*, 1990, **90**, 1079.

The reaction cycle of isopenicillin N synthase observed by X-ray diffraction, N.J. Burzlaff, P.J. Rutledge, J.J. Clifton, C.M.H. Hensgens, M. Pickford, R.M. Adlington, P.L. Roach and J.E. Baldwin, *Nature (London)*, 1999, **401**, 721.

Chemistry and biosynthesis of clavulanic acid and other clavans, K.H. Baggaley, A. Brown and C.J. Schofield, *Nat. Prod. Rep.*, 1997, **14**, 309.

## Section 3.5

The structure of mycelianamide, A.J. Birch, R.A. Massy-Westropp and R.W. Rickards, *J. Chem. Soc.*, 1956, 3717.

Studies in relation to biosynthesis; origin of the terpenoid structures in mycelianamide and mycophenolic acid, A.J. Birch, R.J. English, R.A. Massy-Westropp and H. Smith, *J. Chem. Soc.*, 1958, 369.

The structure of gliotoxin, M.R. Bell, J.R. Johnson, B.S. Wildi and R.B. Woodward, *J. Am. Chem. Soc*, 1958, **80**, 1001.

New piperazinedione metabolites of *Gliocladium deliquescens*, J.R. Hanson and M.A. O'Leary, *J. Chem. Soc., Perkin Trans. 1*, 1981, 218.

Cyclo-(L-phenylalanyl-L-seryl) as an intermediate in the biosynthesis of glio-toxin, G.W. Kirby, G.L. Patrick and D.J. Robins, *J. Chem. Soc., Perkin Trans. 1*, 1978, 1336.

## Section 3.6

Alkaloid biosynthesis in *Penicillium cyclopium*, M. Luckner, *J. Nat. Prod.*, 1980, **43**, 21.

## Section 3.7

Structure and biosynthesis of hinnuliquinone – a pigment from *Nodulisporium hinnuleum*, M.A. O'Leary, J.R. Hanson and B.L. Yeoh, *J. Chem. Soc., Perkin Trans.*, 1984, 567.

## Section 3.8

Agaritine, isolation, degradation and synthesis, R.B. Kelly, E.G. Daniels and J.W. Hinman, *J. Org. Chem.*, 1962, **27**, 3229.
The biosynthesis and possible function of glutaminyl-4-hydroxybenzene in *Agaricus bisporus*, H. Stussi and D.M. Rast, *Phytochemistry*, 1981, **20**, 2347.

## Section 3.9

Cyclosporins – new analogues by precursor directed biosynthesis, R. Traber, H. Hofmann and H. Kobel, *J. Antibiot.*, 1989, **42**, 591.

# Chapter 4

## Section 4.2

Derivatives of the multiple ketene group, J.N. Collie, *J. Chem. Soc.*, 1907, **91**, 1806.
Studies in relation to biosynthesis, 2-hydroxy-6-methylbenzoic acid in *Penicillium griseofulvum*, A.J. Birch, R.A. Massy-Westropp and C.J. Moye, *Aust. J. Chem.*, 1956, **9**, 539.
Biosynthesis of polyketides, T.J. Simpson, *Nat. Prod. Rep.*, 1984, **1**, 28; 1985, **2**, 321; 1987, **4**, 339; 1991, **8**, 573; D. O'Hagan, *Nat. Prod. Rep.*, 1992, **9**, 447; 1993, **10**, 593; 1995, **12**, 1; B.J. Rawlings, *Nat. Prod. Rep.*, 1997, **14**, 335; 1997, **14**, 523; 1998, **15**, 275; 1999, **16**, 425.
Polyketide biosynthesis, a millennium review, J. Staunton and K.J. Weissman, *Nat. Prod. Rep.*, 2001, **18**, 380.
The biosynthesis, molecular genetics and enzymology of polyketide derived metabolites, A.M. Hill, *Nat. Prod. Rep.*, 2006, **23**, 256.

## Section 4.4

Gladiolic acid, a metabolite of *Penicillium gladioli*, J.F. Grove, *Biochem. J.*, 1952, **50**, 648.

Biosynthesis of dihydrogladiolic acid by *Penicillium gladioli*, A.G. Avent, J.R. Hanson and A. Truneh, *Phytochemistry*, 1990, **29**, 2133.

Biosynthesis of fungal metabolites, Terrein, a metabolite of *Aspergillus terreus*, R.A. Hill, R.H. Carter and J. Staunton, *J. Chem. Soc., Perkin Trans. 1*, 1981, 2570.

Biosynthesis of citrinin by *Penicillium citrinum*, J. Barber, R.H. Carter, M.J. Garson and J. Staunton, *J. Chem. Soc., Perkin Trans. 1*, 1981, 2577.

Biosynthesis of fungal metabolites, asperlactone and its relationship to other metabolites of *Aspergillus melleus*, R.G. Brereton, M.J. Garson and J. Staunton, *J. Chem. Soc., Perkin Trans. 1*, 1984, 1027.

Biosynthesis of mycophenolic acid, L. Canonica, W. Kroszczynski, B.M. Renzi, B. Rindone, E. Santaniello and C. Scolastico, *J. Chem. Soc., Perkin Trans. 1*, 1972, 2639.

Mycophenolic acid, A one hundred year odyssey from antibiotic to immuno-suppressant, R. Bentley, *Chem. Rev.*, 2000, **100**, 3801.

## Section 4.6

Griseofulvin, J.F. Grove, *Quart. Rev.*, 1963, **17**, 1.

Biosynthesis of griseofulvin, C.M. Harris, J.S. Robertson and T.M. Harris, *J. Am. Chem., Soc.*, 1976, **98**, 5386.

Biosynthetic source of oxygen in griseofulvin, M.P. Lane, T.T. Nakeshima and J.C. Vederas, *J. Am. Chem. Soc.* 1982, **104**, 913.

## Section 4.7

Microbial pyran-2-ones and dihydropyran-2-ones, J.M. Dickinson, *Nat. Prod. Rep.*, 1993, **10**, 71.

Decanolides, 10-membered lactones of natural origin, C. Drager, R. Kirschning, R. Thiericke and M. Zerlin, *Nat. Prod. Rep.*, 1996, **13**, 365.

$^{13}$C NMR spectrum and biosynthesis of colletodiol, M.W. Lunnon and J. MacMillan, *J. Chem. Soc., Perkin Trans. 1*, 1976, 184.

Biosynthesis of colletodiol and related polyketide macrodiolides in *Cytospora* sp. ATCC 20502, J.A. O'Neill, T.J. Simpson and C.L. Willis, *Chem. Commun.*, 1993, 738.

Bartanol and bartallol, novel macrodiolides from *Cytospora* sp. ATCC 20502, K. Hanson, J.A. O'Neill, T.J. Simpson and C.L. Willis, *J. Chem. Soc., Perkin Trans. 1*, 1994, 2493.

## Section 4.8

Enzymic catalysis of the Diels–Alder reaction in the biosynthesis of natural products, H. Oikawa and T. Tokiwano, *Nat. Prod. Rep.*, 2004, **21**, 321.

Biosynthesis of the hypocholesterolemic agent, mevinolin by *Aspergillus terreus*, R.N. Moore, G. Bigam, J.K. Chan, A.M. Hogg, T.T. Nakashima and J.C. Vederas, *J. Am. Chem. Soc.*, 1985, **107**, 3694; see also *J. Am. Chem. Soc*, 1994, **116**, 2693.

Transformations of cyclic nonaketides by *Aspergillus terreus* mutants blocked for lovastatin biosynthesis, J.L. Sorensen, K. Auclair, J. Kennedy, C.R. Hutchinson and J.C. Vederas, *Org. Biomol. Chem.*, 2003, **1**, 50.

## Section 4.11

Natural polyacetylenes and their precursors, E.R.H. Jones, *Chem. Br.*, 1966, 6.

Natural polyacetylenes part XLII, Novel $C_7$, $C_3$, $C_9$ and $C_{10}$ polyacetylenes from fungal cultures, M.T.W. Hearn Sir Ewart R.H. Jones, M.G. Pellatt, V. Thaller and J.L. Turner, *J. Chem. Soc., Perkin Trans. 1*, 1973, 2785.

Further polyacetylenes from *Polyporus anthracophilus:* Specific incorporation of matricaria esters into polyacetylenic metabolites of this fungus, D.G. Davies, P. Hodge, P. Yates and M.J. Wright, *J. Chem. Soc., Perkin Trans. 1*, 1978, 1602.

# Chapter 5

## Section 5.2

Diversity of the biosynthesis of the isoprene unit, T. Kuzuyama and H. Seto, *Nat. Prod. Rep.*, 2003, **20**, 171.

The biosynthesis of $C_5$–$C_{20}$ terpenoid compounds, P.M. Dewick, *Nat. Prod. Rep.*, 2002, **19**, 181 and previous articles in this series.

Terpenoid biosynthesis: stereochemistry of the cyclization of allylic pyrophosphates, D.E. Cane, *Acc. Chem. Res.*, 1985, **18**, 220.

## Section 5.4

Enzymatic formation of sesquiterpenes, D.E. Cane, *Chem. Rev.*, 1990, **90**, 1089.

Bio-active sesquiterpenes produced by fungi, W.-R. Abraham, *Curr. Med. Chem.*, 2001, **8**, 583.

Non-macrocyclic trichothecenes, J.F. Grove, *Nat. Prod. Rep.*, 1988, **5**, 187;

Macrocyclic trichothecenes, J.F .Grove, *Nat. Prod. Rep.*, 1993, **10**, 429.

Stereochemistry of some stages in trichothecene biosynthesis, R. Evans, J.R. Hanson and T. Marten, *J. Chem. Soc., Perkin Trans, 1*, 1974, 857.

Revision of the stereochemistry in trichodiol, trichotriol and related compounds and concerning their role in the biosynthesis of the trichothecene mycotoxins, A.R. Heskett, B.W. Bycroft, P.M. Dewick and J. Gilbert, *Phytochemistry*, 1993, **32**, 105.

The biosynthesis of some sesquiterpenoids, J.R. Hanson, *Pure Appl. Chem.*, 1981, **53**, 1155.

*Botrytis* species: an intriguing source of metabolites with a wide range of biological activities. Structure, chemistry and bioactivity of metabolites isolated from *Botrytis* species. I.G. Collado, J. Aleu, R. Hernandez-Galan and R. Duran-Patron, *Curr. Org. Chem.*, 2000, **4**, 1261.

Terpenoid metabolites of mushrooms and related Basidiomycetes, W.A. Ayer and L.M. Browne, *Tetrahedron*, 1981, **37**, 2199.

Sesquiterpene aryl esters from *Armillaria mellea*, D.M.R. Donnelly, R.M. Hutchinson, D. Coveney and M. Yonemitsu, *Phytochemistry*, 1990, **29**, 2569.

The sesquiterpenes of *Lactarius vellereus* and their role in a proposed chemical defense system, O. Sterner, R. Bergman, J. Kihlberg and B. Wickberg, *J. Nat. Prod.*, 1985, **48**, 279.

Sesquiterpenes of *Lactarius* origin, anti-feedant structure:activity relationships, W.M. Daniewski, M. Gumulka, D. Przesmycka, K. Praszynska, E. Bloszyk and B. Drozda, *Phytochemistry*, 1995, **38**, 1161.

Conversion of velutinal esters in the fruit bodies of *Russula cuprea*, M. Clericuzio and O. Sterner, *Phytochemistry*, 1997, **45**, 1569.

Biosynthesis of xenovulene A®, M.E. Raggatt, T.J. Simpson and M.L. Chicarelli-Robinson, *Chem. Commun.*, 1997, 2245.

Two new azulene pigments from the fruiting bodies of the Basidiomycete, *Lactarius deliciosus*, X.L. Wang, D.Q. Luo, Z.J. Dong and J.K. Liu, *Helv. Chim. Acta*, 2006, **89**, 988.

Stabilisation of transition states prior to and following eudesmane cation in aristolochene synthase, S. Forcat and R.K. Allemann, *Org. Biomol. Chem.*, 2006, **4**, 2563.

The biosynthetic pathway to abscisic acid via ionylidene-ethene in the fungus, *Botrytis cinerea*, M. Inomata, N. Hirai, R. Yoshida and H. Ohigashi, *Phytochemistry*, 2004, **65**, 2667.

## Section 5.5

Biosynthesis of rosenonolactone, B. Achilladelis and J.R. Hanson, *J. Chem. Soc. (C)*, 1969, 2010.

Application of deuterium magnetic resonance to biosynthetic studies: Rosenonolactone biosynthesis and stereochemistry of a biological $S_N2'$ reaction, D.E. Cane and P.P.N. Murphy, *J. Am. Chem. Soc.*, 1977, **99**, 8327.

Aspects of diterpenoid and gibberellin biosynthesis in *Gibberella fujikuroi*, J.R. Hanson, *Biochem. Soc., Trans.*, 1983, **11**, 522.

The biosynthesis of the gibberellins, J. MacMillan, *Nat. Prod. Rep.*, 1997, **14**, 221.

The chemistry of the gibberellins, J.R. Hanson, *Nat. Prod. Rep.*, 1990, **7**, 41.

Twenty years of gibberellin research, L.N. Mander, *Nat. Prod. Rep.*, 2003, **20**, 49.

Distribution of gibberellin biosynthetic genes and gibberellin production in the *Gibberella fujikuroi* species complex, S. Malonek, C. Bomke, E. Bornberg-Bauer, M.C. Rojas, P. Hedden, P. Hopkins and B. Tudzynski, *Phytochemistry*, 2005, **66**, 1296.

Kaurenolides and fujenoic acids are side products of the gibberellin $P_{450}$-1 mono-oxygenases in *Gibberella fujikuroi*, M.C. Rojas, O. Urrutia, C. Cruz, P. Gaskin, B. Tudzynski and P. Hedden, *Phytochemistry*, 2004, **65**, 821.

Biochemical and molecular analyses in gibberellin biosynthesis in fungi, H. Kawaide, *Biosci. Biotechnol. Biochem.*, 2006, **70**, 583.

Probing the mechanism of loss of carbon-20 in gibberellin biosynthesis, J.L. Ward, P. Gaskin, R.G.S. Brown, G.S. Jackson, P. Hedden, A.L. Phillips, C.L. Willis and M.H. Beale, *J. Chem. Soc., Perkin Trans 1*, 2002, 232.

Biosynthetic sequences leading to the diterpenoid aphidicolin in *Cephalosporium aphidicola*, M.J. Ackland, J.F. Gordon, J.R. Hanson, B.L. Yeoh and A.H. Ratcliffe, *J. Chem. Soc., Perkin Trans. 1*, 1988, 1477.

Biosynthesis of the diterpenoid aphidicolin, Isolation of intermediates from P450 inhibitor treated mycelia of *Phoma betae*, H. Oikawa, S. Ohashi, A. Ichihara and S. Sakamura, *Tetrahedron*, 1999, **55**, 7541.

The constitution of fusicoccin, K.D. Barrow, D.H.R. Barton, E. Chain, U.F.W. Ohnsorge and R. Thomas, *J. Chem. Soc. (C)*, 1971, 1265.

## Section 5.6

The sesterterpenoids, J.R. Hanson, *Nat. Prod. Rep.*, 1996, **13**, 529.

## Section 5.7

Biosynthesis of triterpenoids and steroids, G.D. Brown, *Nat. Prod. Rep.*, 1998, **15**, 653.

Pharmacological activities of natural triterpenoids and their therapeutic implications, P. Dzubak, M. Hajduch, D. Vydra, A. Hustova, M. Kvasnica, D. Biedermann, L. Markova, M. Urban and J. Sarek, *Nat. Prod. Rep.*, 2006, **23**, 394.

Viridin family of steroidal antibiotics, J.R. Hanson, *Nat. Prod. Rep.*, 1995, **12**, 381.

The stereochemistry of fusidic acid, W.O. Godtfredsen, W. von Daehne, S. Vandgedal, A. Marquet, D. Arigoni and A. Melera, *Tetrahedron*, 1965, **21**, 3505.

*Ganoderma* – a therapeutic fungal biofactory, R.R.M. Paterson, *Phytochemistry*, 2006, **54**, 603.

## Section 5.8

Meroterpenoids with various biological activities produced by fungi, K. Shiomi, H. Tomoda, K. Otoguro and S. Omura, *Pure Appl. Chem.*, 1999, **71**, 1059.

Biosynthesis of polyketide-terpenoid (meroterpenoid) metabolites, andibenin A and andilesin A in *Aspergillus variecolor*, T.J. Simpson, S.A. Ahmed, R. McIntyre, F.E. Scott and I.H. Sadler, *Tetrahedron*, 1997, **53**, 4013.

# Chapter 6

## Section 6.3

Revision of the structure of naturally occurring acyltetronic acids; dehydrocarolic acid, terrestric acid and carlic acid, J.P. Jacobsen, T. Reffstrup, R.E. Cox, J.S.E. Holker and P.M. Boll, *Tetrahedron Lett.* 1978, **19**, 1021.

Biosynthesis of carolic acid in *Penicillium charlesii*, J.L. Bloomer, F.E. Kappler and G.N. Pandey, *Chem. Commun.*, 1972, 242.

Total synthesis and absolute configuration of (S)-carlosic acid, J.L. Bloomer and F.E. Kappler, *J. Chem. Soc., Perkin Trans. 1*, 1976, 1485.

## Section 6.4

Two new mould metabolites related to avenaciolide, D.C. Aldridge and W.B. Turner, *J. Chem. Soc. (C)*, 1971, 2431.

Stereoselective synthesis of (+)- and (–)-avenaciolide from D-glucose; the correct absolute configuration of natural avenaciolide, H. Ohrui and S. Emoto, *Tetrahedron Lett.*, 1975, **16**, 3657.

An asymmetric synthesis of naturally occurring canadensolide, R.C. Anderson and B. Fraser-Reid, *Tetrahedron Lett.*, 1978, **19**, 3233.

## Section 6.5

The nonadrides, part 1 Introduction and general survey, D.H.R. Barton and J.K. Sutherland, *J. Chem. Soc.*, 1965, 1769 and following papers.

The nonadrides part V, the biosynthesis of glauconic acid, J.L. Bloomer, C.E. Moppett and J.K. Sutherland, *J. Chem. Soc. (C)*, 1968, 588.

The structure of heveadride, a new nonadride from *Helminthosporium hevea*, R.I. Crane, P. Hedden, J. MacMillan and W.B. Turner, *J. Chem. Soc., Perkin Trans. 1*, 1973, 194.

Rubratoxins, M.O. Moss, F.V. Robinson and A.B. Wood, *J. Chem. Soc. (C)*, 1971, 619.

Natural products with a maleic anhydride structure, nonadrides, tautomycin, chaetomellic anhydride and other compounds, X. Chen, Y. Zheng and Y. Shen, *Chem. Rev.*, 2007, **107**, 1777.

# Chapter 7

## Section 7.1

Pigments of fungi (macromycetes), M. Gill and W. Steglich, *Prog. Chem. Org. Nat. Prod.*, 1987, **51**, 1; *Nat. Prod. Rep.*, 1994, **11**, 67; 1996, **13**, 513; 1999, **16**, 301; 2003, **20**, 615.

The biosynthesis of pigments in Basidiomycetes, M. Gill, *Aust. J. Chem.*, 2001, **54**, 721.

## Section 7.2

Helminthosporin and hydroxyhelminthosporin, metabolic products of the plant pathogen, *Helminthosporium gramineum*, J.H.V. Charles, H. Raistrick, R. Robinson and A.R. Todd, *Biochem. J.*, 1933, **27**, 499.

Crystalline colouring matters of species of the *Aspergillus glaucus* series, B.S. Gould and H. Raistrick, *Biochem. J.*, 1934, **28**, 1640.

Isolation and characterization of a fungal vacuolation factor (Bikhaverin), J.W. Cornforth, G. Ryback, P.M. Robinson and D. Park, *J. Chem. Soc, (C)*, 1971, 2786.

Hypoxylerone, a novel green pigment from the fungus, *Hypoxylon fragiforme*, R.L. Edwards, V. Fawcett, D.J. Maitland, R. Nettleton, L. Shields and A.J.S. Whalley, *Chem. Commun.*, 1991, 1009.

## Section 7.3

Natural terphenyls: developments since 1877, J.-K. Liu, *Chem. Rev.*, 2006, **106**, 2209.

Involutin, a diphenylcyclopenteneone from *Paxillus involutus*, R.L. Edwards, G.C. Elsworthy and N. Kale, *J. Chem. Soc. (C)*, 1967, 405.

An iron(II)-catechol complex as a mushroom pigment, F. von Nussbaum, P. Spiteller, M. Ruth, W. Steglich, G. Wanner, B. Gamblin, L. Stievano and F.E. Wagner, *Angew. Chem. Int. Ed.*, 1998, **37**, 3292.

## Section 7.4

2,3- and 4,5-secoDOPA, the biosynthetic intermediates generated from L-DOPA by an enzyme system extracted from the fly Agaric, *Amanita muscaria* and their spontaneous conversion to muscaflavin and betalamic acid respectively and betalains, F. Terradas and H. Wyler, *Helv. Chim. Acta*, 1991, **74**, 124.

The secoDOPA natural pigments in *Hygrocybe conica* and *Amanita muscaria*, F. Terradas and H. Wyler, *Phytochemistry*, 1991, **30**, 325.

## Section 7.5

The biosynthesis of carotenoids, D.M. Harrison, *Nat. Prod. Rep.*, 1983, **3**, 205.

The apocarotenoid system of sex hormones and prohormones in *Mucorales*, J.D. Bu'Lock, B.E. Jones and N. Winskill, *Pure Appl. Chem.*, 1976, **47**, 191.

## Section 7.6

Biosynthesis of lichen substances, products of a symbiotic association, K. Mosbach, *Angew. Chem. Int. Ed. Engl.*, 1969, **8**, 240.

## Section 7.7

Mushroom flavor, J.A. Maga, *J. Agric. Food Chem.*, 1981, **29**, 1.

Volatile compounds from the mycelium of the mushroom, *Agaricus bisporus*, J.F. Grove, *Phytochemistry*, 1981, **20**, 2021.

Monoterpenes in the aromas of fresh wild mushrooms, S. Breheret, T. Talou, S. Rapior and J.M. Bessiere, *J. Agric. Food Chem.*, 1997, **45**, 831.

Biological activities of volatile fungal metabolites, S.A. Hutchinson, *Annu. Rev. Phytopathol.*, 1973, **11**, 223.

Volatiles of bracket fungi, *Fomitopsis pinicola* and *Fomes fomentarius* and their function as insect attractants, J. Faldt, M. Jonsell, G. Norlander and A.-K Burg-Karlson, *J. Chem. Ecol.*, 1999, **25**, 567.

The secret of truffles; a steroidal pheromone?, R. Claus, H.O. Hoppen and H. Karg, *Experientia*, 1981, **37**, 1178.

Biosynthesis of 13-hydroperoxylinoleate, 10-oxo-8-decenoic acid and 1-octen-3-ol from linoleic acid by a mycelial pellet homogenate of *Pleurotus pulmonarius*, S. Assaf, Y. Hadar and C.G. Dosoretz, *J. Agric. Food Chem.*, 1995, **43**, 2173.

Chemistry in the salad bowl; comparative organosulfur chemistry of garlic, onion and shiitake mushrooms, E. Block and R. Deorazio, *Pure Appl. Chem.*, 1994, **66**, 2205.

A scent of therapy; pharmacological implications of natural products containing redox-active sulfur atoms, C. Jacob, *Nat. Prod. Rep.*, 2006, **23**, 851.

# Chapter 8

Additional references relevant to Chapters 8 and 9 may be found under those for Chapters 3–6.

## Section 8.2

Phytoalexins, C.J.W. Brooks and D.G. Watson, *Nat. Prod. Rep.*, 1985, **2**, 393.

Phytotoxins produced by microbial plant pathogens, R.N. Strange, *Nat. Prod. Rep.*, 2007, **24**, 127.

Endophytes: a rich source of functional metabolites, R.X. Tan and W.X. Zou, *Nat. Prod. Rep.*, 2001, **18**, 448.

Biology and chemistry of endophytes, H.W. Zhang, Y.C. Song and R.X. Tan, *Nat. Prod. Rep.*, 2006, **23**, 753.

Fungal endophytes and their role in plant protection, C. Gimenez, R. Cabrera, M. Reina and A. Gonzalez-Coloma, *Curr. Org. Chem.*, 2007, **11**, 707.

Metabolism and detoxification of phytoalexins and analogs by phytopathogenic fungi, M.S.C. Pedras and P.W.K. Ahiahonu, *Phytochemistry*, 2005, **66**, 391.

Strobilurins; evolution of a new class of active substances, H. Sauter, W. Steglich and T. Anke, *Angew. Chem. Int. Ed.*, 1999, **38**, 1328.

# Section 8.3

Fungal terpene metabolites, biosynthetic relationships and the control of the phytopathogenic fungus, *Botrytis cinerea*, I.G. Collado, A.J. Macias-Sanchez and J.R. Hanson, *Nat. Prod. Rep.*, 2007, **24**, 674.

Botcinins E and F and Botcinolide from *Botrytis cinerea* and structural revision of botcinolides, H. Tani, H. Koshino, E. Sokuno, H.G. Cutler and H. Nakajima, *J. Nat. Prod.*, 2006, **69**, 722.

Dimerization of resveratrol by the grapevine pathogen, *Botrytis cinerea*, R.M. Cichewicz, S.A. Kouzi and M.T. Hamann, *J. Nat. Prod.*, 2000, **63**, 29.

Secondary metabolites isolated from *Colletotrichum* species, C.M. Garcia-Pajom and I.G. Collado, *Nat. Prod. Rep.*, 2003, **20**, 426.

# Section 8.4

The isolation of avenacins A-l, A-2, B-l and B-2, chemical defences against cereal 'take-all' disease. Structure of the aglycones, the avenestergenins and their anhydro dimers, M.J. Begley, L. Crombie, W.M.L. Crombie and D.A. Whiting, *J. Chem. Soc., Perkin Trans 1*, 1986, 1905.

Pathogenicity of 'take-all' fungus to oats; its relationship to the concentration and detoxification of the four avenacins, W.M.L. Crombie, L. Crombie, J.B. Green and J.A. Lucas, *Phytochemistry*, 1986, **25**, 2075.

Hydroxamic acids (4-hydroxy-l,4-benzoxazin-3-ones), defence chemicals in the *Gramineae*, H.M. Niemeyer, *Phytochemistry*, 1998, **27**, 3349.

Detoxification of benzoxazolinone allelochemicals from wheat by *Gaeumannomyces graminis* var. *tritici*, A. Frisbe, V. Vilich, L. Hennig, M. Klugs and D. Sicker, *Appl. Environ. Microbiol.*, 1998, **64**, 2386.

HC-Toxin, J.D. Walton, *Phytochemistry*, 2006, **67**, 1406.

# Section 8.6

Phenolic metabolites of *Ceratocystis ulmi*, N. Claydon, J.F. Grove and M. Hosken, *Phytochemistry*. 1976, **13**, 2567.

Elm bark beetle boring and feeding deterrents from *Phomopsis oblonga*, N. Claydon, J.F. Grove and M. Pople, *Phytochemistry*. 1985, **24**, 937.

Seiricardines B and C, phytotoxic sesquiterpenes from three species of *Seiridium* pathogenic for cypresses, A. Evidente, A. Motta and L. Sparapano, *Phytochemistry*, 1993, **33**, 69.

7′-Hydroxyseiridin and 7′-hydroxyisoseiridin, two new phytotoxic butenolides from three species of *Seiridium* pathogenic to cypresses, A. Evidente and L. Sparapano, *J. Nat. Prod.*, 1994, **57**, 1720.

# Section 8.7

Chemistry of the biocontrol agent, *Trichoderma harzianum*, J.R. Hanson, *Sci. Progr.*, 2005, **88**, 237.

# Chapter 9

## Section 9.1

Mycotoxins, general view, chemistry and structure, P.S. Steyn, *Toxicol. Lett.*, 1995, **82**, 843.

## Section 9.2

Biosynthesis of ergot alkaloids, H. Floss, *Tetrahedron*, 1976, **32**, 873.
From ergot to ansamycins, H. Floss, *J. Nat. Prod.*, 2006, **69**, 158.

## Section 9.3

Mode of action of trichothecenes, Y. Ueno, *Pure Appl. Chem.*, 1977, **49**, 1737.

## Section 9.4

Fusarin C biosynthesis in *Fusarium monoliforme* and *F. venenatum*, Z. Song, R.J. Cox, C.M. Lazarus and T.J. Simpson, *ChemBioChem.*, 2004, **5**, 1196.

## Section 9.5

Metabolism of aflatoxins and other mycotoxins in relation to their toxicity and the accumulation of residues in animal tissue, D.S.P. Patterson, *Pure Appl. Chem.* 1977, **49**, 1723.
Enzymology and molecular biology of aflatoxin biosynthesis, R.E. Minto and C.A. Townsend, *Chem. Rev.*, 1997, **97**, 2537.

## Section 9.6

Production of metabolites from the *Penicillium roqueforti* complex, K.F. Nielsen, M.W. Sumarah, J.C. Frisvad and J.D. Miller, *J. Agric. Food Chem.*, 2006, **54**, 3756.
*Penicillium expansum:* consistent production of patulin, chaetoglobosins and other secondary metabolites and their natural occurrence in fruit products, B. Andersen, J. Smedsgaard and J.C. Frisvad, *J. Agric. Food Chem.*, 2004, **52**, 2421.
Mechanism of patulin toxicity under conditions that inhibit yeast growth, J.Y. Iwahashi, H. Hosoda, J.-H. Park, J.-H, Lee, Y. Suzuki, E. Kitagawa, S.M. Murata, N.-S. Jwa, M.B. Gu and H. Iwahashi, *J. Agric. Food Chem.*, 2006, **54**, 1936.

## Section 9.7

*Mushrooms, Poisons and Panaceas*, D.R. Benjamin, W.H. Freeman and Co., New York, 1995.

On the occurrence of N-methyl-N-formylhydrazones in fresh and processed false morel, *Gyromitra esculenta*, M. Pyysal and A. Niskanan, *J. Agric. Food Chem.*, 1977, **25**, 644.

Isolation and structure of coprine, the *in vivo* aldehyde dehydrogenase inhibitor of *Coprinus atramentarius*, P. Lindberg, R. Bergman and B. Wickberg, *J. Chem. Soc., Perkin Trans. 1*, 1977, 684.

Occurrence of the fungal toxin, orellanine, as a diglucoside and investigation of its biosynthesis, P. Spiteller, M. Spiteller and W. Steglich, *Angew. Chem. Int. Ed.*, 2003, **42**, 2864.

# Chapter 10

## Section 10.1

*Biotransformations in Preparative Organic Chemistry*, H.G. Davies, R.H. Green, D.R. Kelly and S.M. Roberts, Academic Press, London, 1989.

*Biotransformations in Organic Chemistry*, K. Faber, Springer-Verlag, Heidelberg, 1992.

*An Introduction to Biotransformations in Organic Chemistry*, J.R. Hanson, W.H. Freeman, Oxford, 1995.

Recent advances in the use of enzyme catalysed reactions in organic synthesis, N.J. Turner, *Nat. Prod. Rep.*, 1994, **11**, 1.

## Section 10.2

The mechanism of the microbiological hydroxylation of steroids, H.I. Holland, *Chem. Soc. Rev.*, 1983, 371.

Current trends in steroid microbial biotransformations, S.B. Mahato and I. Majumdar, *Phytochemistry*, 1993, **34**, 883 and previous articles.

Bioconversion of sesquiterpenes, V. Lemare and R. Furstoss, *Tetrahedron*, 1990, **46**, 4104.

The biotransformation of sesquiterpenoids by *Mucor plumbeus*, S.F. Arantes and J.R. Hanson, *Curr. Org. Chem.*, 2007, **11**, 657.

## Section 10.3

Biological variation of microbial metabolites by precursor directed bio-synthesis, R. Thiericke and J. Rohr, *Nat. Prod. Rep.*, 1993, **10**, 265.

Microbiological transformation of diterpenoids, J.R. Hanson, *Nat. Prod. Rep.*, 1992, **9**, 139.

The microbiological transformation of two 15β-hydroxy-ent-kaurene diterpenes by *Gibberella fujikuroi*, B.M. Fraga, G. Guillermo and M.G. Hernandez, *J. Nat. Prod.*, 2004, **67**, 64.

# Glossary

**Actinomycetes**
A group of mycelium-forming bacteria of which the *Streptomyces* is a common genus.

**Agar**
A polysaccharide obtained from sea-weed which forms a gel. When this is supplemented with nutrients, it is used as a support medium for growing fungi.

**Agaric**
A fungus that forms a mushroom-like fruiting body. These are umbrella-shaped and may contain gills. The name may refer more specifically to organisms that belong to the family Agaricaceae.

**Ascomycetes**
A sub-division of the fungi in which the spores (the ascospores) are contained within a small structure like a bag, which is known as an ascus.

**Bacteria**
Prokaryotic microorganisms that are distinct from fungi, which are eukaryotic. Prokaryotic organisms lack a true nucleus. Their DNA is present within the cytoplasm. Bacteria are usually unicellular and have a rigid cell wall. Cell division usually occurs by binary fission.

**Basidiomycetes**
A sub-division of the fungi in which the spores (basidiospores) are produced on distinct fruiting bodies.

**Chitin**
A polysaccharide containing $N$-acetyl-D-glucosamine, which is a major constituent of the fungal cell wall. It is also a major component of the cuticle of an insect.

**Conidiophore**
A specialized hyphae on which the conidia (spores) are borne.

**Cyanobacterium**
A group of photosynthetic bacteria that are sometimes known as the blue-green algae. Some are found growing in association with fungi in lichens.

**Czapek–Dox medium**
A defined synthetic medium for growing fungi. Its constitution is given in Chapter 2.

**Deuteromycetes**
A sub-division of the fungi containing the Fungi Imperfecti in which a perfect stage of reproduction is unknown or rarely found.

**Endophytic organisms**
Organisms, particularly fungi, found growing entirely within a plant.

**Entomogenous fungi**
Fungi that grow on insects.

**Entomophilous fungi**
Fungi whose spores are distributed by insects.

**Eukaryote**
An organism whose cells have a distinct nucleus. These include the fungi, plants and animals.

**Fruiting body**
The differentiated spore-bearing structure such as the ascocarp or basidiocarp that is a characteristic of the higher fungi.

**Fungi**
Eukaryotic, non-photosynthetic organisms that obtain their nutrients by the absorption of compounds from their surroundings.

**Fungi Imperfecti**
Another name for the Deuteromycetes.

**Fungicolous**
An organism that grows on a fungus.

**Gasteromycetes**
A group of Basidiomycetes with well-established fruiting bodies.

**Hymenium**
A layer of spore-bearing structures typically found in polypores such as the bracket fungi growing on decaying wood.

**Hyphae**
Thread-like filaments that form the basic structural units in fungi.

**Idiophase**
The growth phase of an organism in which it forms its characteristic metabolites.

**Lag phase**
The stage of fungal growth in which an organism is becoming established on the growth medium before the rapid exponential growth phase begins.

**Lichen**
A composite organism that consists of a fungus growing in a symbiotic relationship with a photosynthetic algae or cyanobacterium.

**Lysis**
The rupture of a cell.

**Macrofungi**
A name given to the larger Ascomycetes and Basidiomycetes. They are sometimes referred to as the higher fungi.

**Microfungi**
A name given to the smaller filamentous fungi.

**Mycelium**
The mat of fungal hyphae that make up the vegetative stage of the growth of a microorganism.

**Mycorrhizal**
The close symbiotic relationship between fungi and the roots of plants in which the plant assimilates the nutrients released from the soil by the fungus. In return the fungi gain some nutrients from the plant.

**Mycotoxin**
A fungal metabolite that is toxic to mammals. The term is often restricted to metabolites that are toxic to man.

**Perfect stage**
The sexual stage of fungal development in which sexually produced spores are formed.

**Perithecium**
A rounded ascocarp from which the spores are discharged *via* a small hole.

**Petri dish**
A shallow circular dish with a lid that is used for incubating microorganisms.

**Phytotoxin**
A compound that is toxic to plants.

**Plasma membrane**
The cell membrane that encloses the contents of the cell. Its structure permits the passage of materials both in and out of the cell. Various enzyme systems may be attached to this structure.

**Polyketides**
Compounds formed by the linear condensation of acetate (or occasionally propionate) units.

**Polypore**
Fungi that possess the hymenium as tubes that open to the exterior as pores. The bracket fungi are examples.

**Prokaryotes**
Organisms, typically a bacterium, in which the cells lack a true nucleus.

**Raulin–Thom medium**
A standard synthetic medium. For contents see Chapter 2.

**Rhizomorph**
A thick differentiated strand of hyphae appearing like a root or 'boot-lace' found, for example, in several Basidiomycetes.

**Roux bottle**
A flat fermentation bottle usually of one-litre total volume with an off-set narrow neck. Typically, it will be used to accommodate 150 mL of medium. A larger bottle is sometimes called a Thompson bottle whilst a circular 'bed-pan' is sometimes known as a Penicillin flat or flask.

**Rust fungi**
A group of obligate fungal parasites on plants belonging to the class Urediniomycetes which produce masses of rust-coloured spores and can cause serious plant diseases.

**Saprophyte**
An organism that grows on dead material.

**Sclerotium**
The resting stage of a fungus that is resistant to unfavourable growing conditions and may be dormant.

**Septate**
Hyphae with partitions dividing it into separate cells. The term aseptate is used for hyphae that lack these divisions.

**Smut fungi**
Members of the Ustilaginaceae that are fungal parasites of plants that form a mass of dark powdery spores and cause disease particularly of cereals.

**Spore**
A microscopic structure that functions in the reproduction and dispersal of fungi rather like a seed.

**Stipe**
A stalk or stem of the fruiting body of a fungus.

*Streptomyces*
A genus of bacteria that contains a well-defined mycelium.

**Symbiosis**
The situation in which dissimilar organisms grow together, often in a mutually beneficial manner.

**Terpenoids**
Compounds biosynthesized from isopentenyl diphosphate in which the $C_5$ units are normally attached in a head-to-tail manner.

**Thallus**
A part of the simpler fungi which may bear the spores.

**Tropophase**
The growth phase in fungal development.

**Vacuole**
A membrane-bound sac found in a cell which can act as a storage organ.

**Volva**
A cup-like membrane as the base of the stem of the fruiting body of a fungus.

**Wild type**
The naturally occurring strain of an organism as opposed to variants resulting from laboratory-based mutant and strain enhancement techniques.

# Subject Index

*Note*: Page numbers in *italic* refer to figures. **Bold** numbers indicate more extensive treatment of a topic.

abscisic acid 153–4
accession number 4, 5
acetyl co-enzyme A 48, *49*, 120, *121*
acetylorsellinic acid 58
*Acremonium strictum* 86
*Acrocylindrium oryzae* 111
Actinomycetes 1, 204
activated carbon 22
aflatoxins 15, **169–71**
agar 204
agaratine 45
agaric 204
agaricic acid 123
agaricone 136
*Agaricus* sp. 28
  *A. bisporus* 2, 4, 44, 45, 142
  *A. xanthoderma* 136
*Agrocybe dura* 70
agrocybin 70, *71*, 92
alcohol sensitization 176
alimentary toxic aleukia 166–7, *168*
alkaloids 7, 184
allelochemicals 148, 156
alliacol A 92
alliacolide 92
altenuic acid 152, *153*
altenusin 152, *153*
*Alternaria* sp. 12, 47, 50
  *A. alternata* 45, 153
  *A. helianthi* 50

*A. kikuchiana* 159
*A. radicina* 50
*A. solani* 151
*A. tenuis* 152
  leaf spot diseases 151–4
alternaric acid 151, *152*
alternariol 152
AM-toxin 45
*Amanita* sp.
  *A. muscaria* 6, 127, 135
  *A. pantherina* 174
  *A. phalloides* 173, 174
  *A. virosa* 173
  poisonous mushrooms 173–5
amanitins 173–4
*Amillaria mellea* 3
amino acids 30, 32
  fungal metabolites from **33–46**
6-aminopenicillanic acid 10–11, 35, 36
amoxicillin 35, 36
amphotericin B 5
ampicillin 36
analogue biosynthesis 183
andibenin A 117, *118*
andilesin A 117, *118*
androst-16-en-3-one 145
5α-androstanes 181, 182
anthracnose diseases 154, 155
anthraquinone pigments 9, 129–30, 132

anti-fungal agents 5, 12, 14, 46, 58,
    61, 64, 111
  and *Botrytus* infection 150–2
  plant–fungal interactions 148
  polyketides 58, 61, 64
  *Trichoderma* sp. 163–4
anti-tumour agent 101
antibiotics 5, 6, 14
  historical review 10–12
antiviral agent 58, 68, 101
aphidicolin 14, **101**
aranotin 42
aristolochene *82*
*Armillaria mellea* 3, 14, 27, 88, **161**
*Armillaria tabescens* 88
armillyl orsellinate 88, 161
Arnstein tripeptide 11, 37, *38*
arylpyruvic acids 132
Ascomycetes 4, 204
*Asochyte rehiei* 152
asperentin 64
aspergillic acid 39, *40*
aspergilliosis 42
*Aspergillus* sp. 4, 5, 9, 27, 47, 58,
    116
  *A. alliaceus* 184
  *A. amstelodami* 39
  *A. avenaceus* 123
  *A. echinulatus* 39
  *A. flavipes* 15, 55
  *A. flavus* 15, 39, 64, 170
  *A. fumigatus* 11, 26, 41, 42, 52, 107,
      117, 121, 128
  *A. giganteus* 33
  *A. glaucus* 129
  *A. melleus* 60
  *A. microcysticus* 68
  *A. niger* 7, 8, 120, 127, 131, 179, 182,
      183
  *A. ochraceus* 172, 180
  *A. oryzae* 8
  *A. repens* 64
  *A. sylvaticus* 42
  *A. terreus* 14, 26, 43, 60, 67, 121
  *A. ustus* 104
  *A. variecolor* 117

asperlactone 60, *61*
aspyrone 60, *61*
astaxanthin 139
atherosclerosis 117
Athlete's Foot 20, 165
atromentin 9, 133
augmentin® 35
aurantricholone 134
aurofusarin 131
auroglaucin 129
avenacins 156, 157
avenaciolide 123–4
azole fungicides 5
azoxystrobin 14

*Bacillus cereus* 36
*Bacillus subtilis* 35
bacteria 204
  culture contamination 20
Bakers' yeast 179
Balkan endemic nephropathy 172
bark beetle 159
bartanol 66
Basidiomycetes 4, 28
  culture medium 18, 19
  defined 204
  polyacetylenes 70
  sesquiterpenoids 85–93
  triterpenoids 113–16
*Beauveria bassiana* 45
*Beauveria sulfurescens* 182
beauvericin 45
beauverolides 45
betalamic acid 135, *136*
bikhaverin 69, 131
bio-control 146, 164
bioluminescence 87–8, 92
biosynthetic studies 16, **29–31**
  historical review 15–17
  metabolite classes/groups 30–1
biotransformations 15, 24–5, **177–87**
  biosynthetically-patterned 183–7
  xenobiotic
    hydrolysis 178
    hydroxylation 180–3
    redox 179–80

*Bipolaris* sp. 155
  *B. maydis* 104
  *B. oryzae* 4, 104
  *B. sorghicola* 104
  *B. sorokinianum* 84, 156
bisabolenes 77
bislactones 66
black rot 50
black spot 26
*Blakeslea trispora* 138
blennione 137
Bordeaux mixture 7, 20
botrallin 152
botryanes 81–4
botrydial 81, *82*, 150
*Botrytis* sp. 149
  *B. allii* 11, 61, 78
  *B. cinerea* 3, 14, 81, 82, 83
    leaf spot disease 149–51
brefeldins 70
byssochlamic acid 124
*Byssochlamys fulva* 124

cadinane 92, *93*
cadmium 28
*Caldariomyces fumago* 26
*Calonectria decora* 180, 181
candensolide 123–4
*Candida* sp.178
  *C. albicans* 5
  *C. cylindracea* 178
  *C. rugosa 179*
canker
  cypress (*C. sempervirens*) 163
  *Nectria galligena* 162–3
*Cantharellus* sp.
  *C. cibarius* 138
  *C. cinnabarinus* 138
  *C. infundibilis* 138
canthaxanthin 138
*Capsicum frutescens* 150
capsidiol 150, 151
cardiac beriberi 171
carlic acid 122
carlosic acid 9, *10*, 122
carolic acid 122

carolinic acid 122
carotenoids 73, **138–40**
catenarin 130
cell wall, fungal 5–6
CellCept® 58
cephalosporins **36–7**
  cephalosporin C 11, 13, 36-7
  cephalosporin $P_1$, 107, *109*, 110
  structures *37*, *109*
*Cephalosporium acremoniun* 11, 36, 39
*Cephalosporium aphidicola* 14, 65, 101, 107
cephalosporolides 65
*Ceratocystis* sp. 76
  *C. coerulescens* 76
  *C. ulmi* 14, 76, 147, **159–60**
*Cercospora rosicola* 153
cereal diseases 4, 12, 153, **155–7**, 159
cesium 28
α-cetylcitric acid 6
chaetoglobosins 68
*Chaetomium cochliodes* 43
*Chaetomium globosum* 68
charcoal 22
cheese 146
cheimonophylion A 77
*Cheimonophyllum candidissimium* 77
Chernobyl accident 28, 134
Chinese medicine 115, 116
chiral reactions 178–80
chitin 204
  synthases 5–6
chloramphenicol 11, 36
chloride ions 25–6
*Chlorosplenium aeruginosum* 131
cinnatriacetins 71
cinnibarinic acid 137
citreoviridin 171, *172*
citric acid 7, *8*, 120–2
  cycle 31, **120–6**
citrinin 9, *10*, 58–9, **58–9**, 172
cladospolide A 65
cladosporin **64–5**

*Cladosporium* sp.
  C. *cladosporioides* 64
  C. *fulvum* 65
  C. *recifei* 65
classification
  fungal metabolites 30–1
  fungi 4–5
*Claviceps paspali* 166
*Claviceps purpurea* 6, 7, 42, 165, 184
clavulanic acid 35, 36
*Clitocybe illudens* 87, 88, 89
*Clitocybe odora* 142
*Clitopilus scyphoides* 102
*Clostridium acetobutylicum* 8
cochliodinol 43, *44*
cochliodinone 43, *44*
colletochlorin B 162, *163*
colletodiol 51, 66, 155
colletoketol 66, 155
colletol 66
colletollol 66
colletorin B 162, *163*
colletotric acid 154, *155*
*Colletotrichum* sp. 154
  C. *capsici* 51, 66, 155
  C. *gloeosporioides* 154
  C. *nicotiniana* 155
colours *see* pigments
compactin 67
conidiophore 3, 204
*Conocybe cyanopus* 175
coprine 176
*Coprinus atramentarius* 176
coriolins 91
*Coriolus consors* 91
corked wine 27
corn steep liquor 20, 33
cortical steroids 15, 180–1
*Corticium caeruleum* 133
*Cotinarius* sp. 132
  C. *violaceus* 127
cotylenins 103
crepenynic acid 71, *72*
culmorin 84–5
culturing and fermentation **18–28**
  contamination 20, 21

culture conditions 3, 20
culture media 18–20
  utilization of constituents 25–7
fermentation
  laboratory techniques 20–1
  stages in 23–5
fungi in the wild 28
metabolite isolation 21–2
*Cupressus sempervirens* 163
curcubitaccins 150
cyanobacterium 204
cyclonerodiol 77
cyclopenin-viridicatin group **42**, *43*
cyclosporins 14, 46, 183–4
*Cylindrocarpon heteronemum* 162
*Cylindrocarpon lucidum* 14
cypress tree 163
cytochalasins 68, *69*
cytochrome P450 oxidases 19
*Cytospora* sp. 66
  C. *capsici* 65
Czapek–Dox medium 18, *19*, 205

damp rooms 167
decarestrictins 65
α,β-dehydrocurvularin 162, *163*
dehydromatricaria acid 71, *72*
demethoxyviridin 111, *112*, 113
depsides 140
depsidones 140
depsones 140
*Dermocybe cinnamomeolutea* 132
*Dermocybe sanguinea* 9
destruxins 45
Deuteromycetes 4, 205
diacetoxyscirpenol 158, 168
diaporthin 159
2,4-dichlorophenoxyacetic acid 183
dieback, Eutypa 160–1
dihydrobotrydial 81, *82*, 150
dihydrogladiolic acid 55, 159
1,8-dihydroxynaphthalene 5, *6*
3,4-dihydroxyphenylalanine 5, *6*
diketopiperazines **39–40**
dilactones 51
dimethyl disulfide 145

*Diplocarpon rosae* 153
*Diplocarpus rosae* 26
*Diplodia pinea* 65
diplodialides 65
*Diploicia canescens* 140
diploicin 140, *141*
diseases, plant *see* plant diseases
diterpenoids 73, **93–103**
   aphidicolin 101–2
   fusicoccins and cotylenis 102–3
   gibberellins and kaurenolides 97–100
   pleuromutilin 102
   rosanes 94–6
   virescenosides 94
2,4-dithiapentane 145
dithiosilvatin 42
L-DOPA 136
drosophilin A methyl ether 143, 144
dry-rot 70–1, 133, 142
Dutch elm disease 14, 76, 147, **159–60**

eburicoic acid 113, *114*
echinulin 39
endophytes 2, 205
*Endothia parasitica* 159
entomogenous fungi 205
entomophilous fungi 205
enzymes
   in biotransformation 183
   enzymatic reduction 179–80
*Epicoccum nigrum* 56
epoxides, microbial hydrolysis 179
ergosterol 5, 7, 105–6, **106–7**
   biosynthesis *108*
ergot alkaloids 184
ergot and ergotism 6, 7, **165–6**
ergotamine 166
ergotinine 7
eritadenine 144
erythoglaucin 129
erythromycin 11, 184
*Escherichia coli* 29, 36, 38
eukaryote 1, 205
eulachromene 160
eutapine 161
Eutypa dieback **160–1**

*Eutypa lata* 160
extraction/isolation techniques 21–2

fairy ring fungus 91
Farmer's Lung 165
farnesyl diphosphate 75–6, 77, 84, *87*,
   105, *106*
fasciculic acid A 115
fatty acids, fungal 47, **68–70**
   stereochemistry of biosynthesis 64
fermentation
   laboratory techniques 20–1
   stages in 23–5
   *see also* culturing
ferriaspergillin 39
fescue foot 167
*Fistulina hepatica* 71
flavipin 18, *19*, 55–6
flavoglaucin 129
Fleming, Alexander 10
fly agaric 6, 127, 135
fomannoxin 160
fomentariol 135
*Fomes annosus* 89, 160
*Fomes fomentarius* 135
*Fomitopsis officinalis* 6
*Formitopsis insularis* 88
frangulaemodin 9
fruiting body 3, 205
Fulcin® 61
fumanosins 69, *70*
fumaric acid 121
fumigatin 11, *12*, 52, **128–9**
fumonisins 168, *169*
fungal metabolites *see* metabolites
fungi
   cell wall 5–6
   classification 4–5
   defined and described 1–2, 205
   metalbolite review 6–13
   structure 2–3
Fungi Imperfecti 4, 205
fungicides *see* anti-fungal agents
fungicolous 205
fungisterol 113
fusarin C 168, *169*

*Fusarium* sp. 4, 47, 80
  *F. avenaceum* 147
  *F. crookwellense* 84
  *F. culmorum* 77, 84, 147
  *F. graminearum* 84, 131, 158, 167,
    168
  *F. monoliforme* 4, 157, 168
  *F. nivale* 167
  *F. oxysporum* 131, 158
  *F. sporotrichioides* 167
  *F. tricinctum* 158
  *F. venenatum* 15
  mycotoxins from 166–9
fusicoccins 102–3
*Fusicoccum amygdali* 102
fusidane steroidal antibiotics 107–11
fusidic acid 109–11
*Fusidium coccineum* 111

*Gaeumannomyces graminis* 157
ganoderic acids 115, *116*
*Ganoderma lucidum* 6, 115
Gasteromycetes 205
genera 4–5
geosmin 146
geranyl diphosphate 75, *76*
geranylfarnesyl diphosphate 104–5
geranylgeranyl diphosphate 93, 99,
  100, 103
*Gibberella fujikuroi* 4, 9, 12, 31, 69, 77
  and biotransformations 184–7
  fermentation 23, 24
  gibberellins/kaurenolides 97, 98, 100
  red pigment 131
gibberellic acid 12, *13*, 16, 23, 24, 184
  biosynthesis 98–9, 184–5
  degradation 97–8
gibberellins 97–100, 185–7
gladiolic acid 13, **55–6**, 158, *159*
glauconic acid 124
*Gliocladium* sp. 40
  *G. deliquescens* 41
  *G. fimbricatum* 11, 41
  *G. roseum* 178
  *G. virens* 111
gliotoxin 10, *11*, 26, *27*, **40–2**

global warming 164
gloeosporone 154
D-gluconic acid lactone 8
glutamic acid derivatives **44–5**
γ-glutamylmarasmane 144
gomophidic acid 133, *134*
Gramineae, diseases of 4, 12, 153,
  **155–7**
grasses 155
grevillin 132
*Grifolia confluens* 116
grifolin 116, *117*
griseofulvin 11–12, 16, 26, **61–4**
Grisovin® 61
growth
  wild fungi 28
  *see also* culturing
growth phase 23, *24*
*Gyromitra esculenta* 175
gyromitrin 175, *176*
gyrophoric acid 140
gyroporin 135
*Gyroporus cyanescens* 135

hallucinogens 6, **174–5**
harzianolide 61, 164
HC-toxin 45, 156
*Hebeloma longicaudum* 87
helicobasidin 78
*Helicobasidium mompa* 78
helminthosporal 84, *85*, 156
helminthosporin 84–5, 129–30
*Helminthosporium* sp. 9, 116, 155
  *H. carborum* 156
  *H. catenarium* 130
  *H. dematioideum* 68
  *H. heveae* 125
  *H. oryzae* 4, 104
  *H. sativum* 84, 85, 156
  *H. siccans* 116
  *H. victoriae* 156
  *H. zizaniae* 104
helminthosporol 84, *85*, 156
helvolic acid 10, *11*, 110–11
heptaketides 61–4
heveadride 125

hinnuliquinone 43, *44*
hirsutic acid 91
hispidin 137
historical review 6
  ninteenth century 7–8
  1900–1940 8–10
  1940–1965 10–13
HMGCoA reductase 66
honey-fungus 3, 14, 27, 88, **161**
HS-toxin 148
humulene 86, *87*
hydrolysis, microbial 178–80
13-hydroperoxy-*cis*-9-*trans*-11-
  octadecadienoic acid 143
2-hydroxy-4-iminocyclohexa-2,5-
  dieneone 44, *45*
hydroxyanthraquinone pigments 129–30
hydroxyaspergillic acid 39, *40*
hydroxylation, microbial 180–3
hymenium 205
hyphae 3, 205, 207
*Hypholoma* sp.
  *H. fasciculare* 115, 137
  *H. sublateritium* 137

ibotenic acid 174, *175*
ice-man 6
idiophase 23, *24*, 205
illudalanes 89
illudins 88–9
illudol 87, *88*
immunosuppressants 8, 42, 46, 58
indolylpyruvic acid 43, *44*
ink-cap mushroom 176
*Inonotus hispidus* 137
involutin 135
islandicin 130
isolation/extraction techniques 21–2
isopenicillin N synthase 37, 38, 39
isopentyl diphosphate 73, 75
  biosynthesis *74*
isoseiridin 163
isotopic labelling 16, 29–30
isovelleral 90
isoviresenol A and B 94
itatartaric acid 121, *122*

Johnson grass 156

kaurenolides 98–100, 186
ketides *see* polyketides

β-lactamase inhibitor 35
β-lactams 10, 11
  biosynthesis 37–9
  cephalosporins 36–7
  penicillins 34–5
lactarazulenes 86
*Lactarius* sp. 85, 90
  *L. blennius* 137
  *L. delicosus* 85
  *L. mitissimus* 86
  *L. piperatus* 90
  *L. rufus* 90
  *L. vellereus* 90
lactaroviolin 85–6, 138
lactones, polyketide 65–6
lag phase 23, *24*, 205
lampterol 88
*Lampteromyces japoicus* 88
lanosterol 106, *107*
lead 28
leaf-spot diseases 149–55
*Lecanora atra* 140
lecanoric acid 140
lentinacin 144
lentinic acid 144
*Lentinus edodes* 144
*Lentinus lepideus* 27, 92
*Letharia vulpina* 141
Liberty Cap 175
lichens 1, 123, 206
  pigments 140–2
lichensterinic acid 141, *142*
Ling Zhi 115
LLD-ACV (tripeptide) 11, 37, *38*
*Lobesia botrana* 149
lovastatin 14, 67
lower fungi 4
*Lupinus luteus* 150
luteone 150, *151*
lysergic acid 7, 166
  diethylamide 166
lysis 206

macrodiolides 66
macrofungi 205
magic mushrooms 175
*Malassezia globosa* 165
malonyl co-enzyme A 48, *49*
marasmane 92
marasmic acid 90
*Marasmius* sp.
    *M. alliaceus* 92, 144
    *M. conigenus* 90
    *M. oreades* 70, 91, 92
*Megasella halterata* 143
melanins 5
melleins 58, 160
melleolide 88, 161
meroterpenoids 57, 73, **116–19**, 155
metabolites, fungal
    classified 30–1
    historical review 6–13
metal ions 19–20, 28
*Metarrhizium anisopliae* 45
methicillin resistance 11
2-methoxy-3-isopropylpyrazine 146
β-methoxyacrylates 14
2-methylisoborneol 146
5-methylmellein 160
6-methylsalicylic acid 16, 17, 25, **51**
    polyketide biosynthesis 48, *49*
L-γ-methyltetronic acid 122
mevalonate pathway 31
mevalonic acid 73–4, *74*
mevalonolactone 178
mollisin 16
*Monascus* sp. 4
    *M. ruber* 4, 67
monoliformin 168, *169*
*Monolinia laxa* 56
monoterpenoids 73, 76
*Morteierella alpina* 69
*Morteierella isabellina* 69
*Mucor* sp. 4, 27
    *M. plumbeus* 182
    *M. stolonifer* 8
multicolic acid 123
muscarine 174, *174*
muscimol 174, *175*

muscoflavin 136
mushrooms 3, 44
    common 2, 4, 44, 45, 142
    poisonous **173–6**
mycelene 40
mycelianamide **40**
mycelium 3, 206
mycophenolic acid 8, **57–8**
mycoprotein 15
mycorrhizal fungi 2, 206
mycosporins 154
mycotoxins 13, 15, 43, **165–76**
    aflatoxins 15, 169–71
    defined 206
    ergotism 6, 7, **165–6**
    *Fusarium* metabolytes 166–9
    poisonous mushrooms 175–6
    polyketides 48
    trichothecines 166–8
*Myrothecium* sp. 80
    *M. roridum* 111

*N*-acetylglucosamine 5
*N*-jasmonyl isoleucine 69, *70*
*Neamatoloma fasciculare* 115, 137
*Nectria galligena* 162
nectriapyrone 73
neoilludol 88
*Neurospora crassa* 97, 138, 139
*Nodulisporium hinnuleum* 43, 111, 113
nonadrides 124–6
norbadione 28
norlichexanthone 141
nystatin 5

ochratoxin A 172
oct-1-en-3-ol 142, *143*, 146
oct-1-en-3-one 142, *143*
octaketides 61–5
odours, fungal **142–6**
    sulfur containing 144–6
Onylalai 153
*Oospora lactis* 69
*Oospora virescens* 94
ophiobolins 4, **104–5**
organoleptic components **142–4**

*Orphiobolus hererostrophus* 104
orsellinic acid 47, 51, *52*, 140
oudemansins 14
oxaloacetic acid 120, *121*
10-oxo-*trans*-8-decenoic acid 143

*Paecilomyces tenuipes* 184
*Paecilomyces variotii* 27
panal 92, *93*
*Panellus stipticus* 92
paspalum staggers 166
Pasteur, Louis 7
patulin 15, **52–5**, 172, 172–3
*Paxillus atromentosus* 9, 133
peach twig blight 56
pellagra 9
penaldic acid 33, *34*
penicillamine 33, *34*
penicillic acid 8, 9, **54–5**, 172
  biosynthesis *55*
penicillins 1, 10–11, **33–6**
  biosynthesis 36–7
  penicillin N 36, *37*
  structure *34*
*Penicillium* sp. 3, 4, 9, 10, 16, 47, 116
  mycotoxins from 171–3
  *P. brefeldianum* 70
  *P. brevicaulis* 27
  *P. brevicompactum* 8, 57, 58, 67, 159
  *P. camembertii* 146, 173
  *P. canadense* 123
  *P. caseicolum* 146
  *P. charlesii* 9, 122
  *P. chrysogenum* 2, 26, 33, 127
  *P. cinerascens* 122
  *P. citreoviride* 171
  *P. citrinum* 9, 67, 172
  *P. crustosum* 173
  *P. cyclopium* 8, 9, 42, 54
  *P. expansum* 53, 172, 173
  *P. gladioli* 55, 158
  *P. glaucum* 7, 124
  *P. griseofulvum* 11, 25, 40, 51, 61
  *P. islandicum* 130
  *P. janczewskii* 12, 61
  *P. luteum* 8

*P. multicolor* 123
*P. nigricans* 39
*P. notatum* 10, 26, 33
*P. patulum* 50, 52–4, 173
*P. puberulum* 9, 54, 56, 172
*P. purpurogenum* 125
*P. roqueforti* 43, 81, 172, 173
*P. rubrum* 10, 125
*P. simplicissimum* 65
*P. solitum* 67
*P. stipitatum* 9, 56
*P. stoloniferum* 8, 57
*P. terlikowski* 41
*P. urticae* 53
*P. wortmanii* 111
penicilloic acid 33, *34*
penillamines 34, *35*
penillic acids 34, *35*
penilloaldehyde 33, *34*
penilloic acid 34
pentaketides 58
  citrinin 58–9
  terrein 60–1
6-n-pentylpyrone 146, 164
peptabiols 45
peptides, fungal **45–6**
perfect stage 206
perithecium 206
*Pertusaria amara* 140
petri dish 206
*Pfaffia rhodozyma* 139
*Phaeosphaeria* sp. 97
phallotoxins 173–4
*Phallus impudicus* 142
*Phellinus* sp. 26
  *P. tremulae* 91
*Phoma* sp. 68, 126
  *P. putaminum* 65
  *P. sorghina* 153
phomins 68
*Phomopsis oblonga* 160
phorid fly 143
*Phycomyces blakesleanus* 138
Phycomycetes 4
physcion 129, 141, *142*
phytoalexins 148, 150

*Phytophthora* sp. 4
  *P. cinnamomi* 162
  *P. infestans* 147
  *P. kernoviae* 162
  *P. ramorum* 162
phytotoxins 13, 104, **148**
  defined 206
  ophiobolins 104
  radicinin 50–1
  *see also* plant diseases
picrolichenic acid 140, *141*
pigments 7, 9, 39, 43, 69, **127–42**
  hydroxyanthraquinones 129–30
  naphthopyrones 130–1
  nitrogen containing 135–8
  pulvinic acids 133–5
  quinones 9, 11, **128–9**
    dimeric and extended 131–2
  terpenoid 138
    carotenoids 138–40
  terphenyls 132–3
  xanthones and naphthopyrones 130–1
pinicolic acid A 113
pipersial 90, *91*
*Piptoporus betulinus* 6
*Piptopterus* sp. 5
*Pisolithus arrhizus* 134
plant diseases 4, 7, **147–64**
  1940–1965 reviewed 12–13
  cereals 4, 12, 153, **155–7**
  chemistry of plant–fugal interactions 148
  and Global Warming 164
  leaf-spot
    *Alternaria* sp. 151–4
    *Botrytis cinerea* 149–51, 154
    *Cercospora* sp. 153–4
    *Colletotrichum* sp. 154
  root infection 157–9, 161, 162
  tree diseases **159–63**
    canker 162–3
    Dutch elm 159–60
    Eutypa dieback 160–1
    Honey fungus 161
    *Phytophthora cinnamomi* 162
    Silver-leaf 162

*Trichoderma* anti-fungals 163–4
Plantsheild® 164
plasma membrane 206
*Plasmopara viticola* 7
pleuromutilin 14, **102**
*Pleurotus mutilus* 14
poisonous mushrooms 173–6
poly-1,3-diketones 47
polyacetylenes 13, **70–2**
polyene antibiotics 5
polyketides 17, 30, **47–72**
  biosynthesis 48–50
    sterochemistry 64–5
  cytochalasins 68, *69*
  defined and described 47–8, 206
  hepta- and octaketides 61–5
  pentaketides 58–61
  pigments 48, **128–32**
  polyketide lactones 65–6
  statins 66–7
  tetraketides 51–8
  triketides 50–1
polypore 206
polyporenic acids 113, *114*, 115
polyporic acid 9, 133
*Polyporus* sp.
  *P. anthracophilus* 113
  *P. betulinus* 113
  *P. enthracophilus* 113
  *P. nidulans* 9, 133
  *P. officinalis* 123
  *P. pinicola* 113
*Populus tremuloides* 91
*Poria cocos* 116
poricoic acid 116
potato blight 147, 151
PR-Toxin 81, *82*
pravastatin 67
precursor-directed biosynthesis 183
Prelog's rule 179–80
progesterone 180, *181*
prokaryotes 1, 207
pruning 160
*Pseudomonas putida* 182
psilocin 175
psilocybin 175

*Psylocybe mexicana* 175
*Psylocybe semilanceata* 175
puberulic acid 9, 56
pulvinic acids 133–5
putaminoxin 65
*Pycnoporus cinnebarinus* 137
pyrenolides 65
Pyrenomycetes 4
*Pyrenophora teres* 65
*Pyricularia oryzae* 159
pyropenes 117
*Pythium* sp. 4
  P. debarynum 147
  P. ultimum 147

quinone pigments 9, 11, **128–9**
  dimeric and extended 131–2
Quorn® 15

radicinin 50–1
radiolabelling 16, 29–30
*Radulomyces confluens* 88
Raulin–Thom medium 18, *19*, 207
reagents, fungi as *see*
    biotransformations
recifeiolide 65
redox reactions, microbial 179–80
reductases 66
Reishi 115
resveratrol 150, *151*
*Rhizoctonia solani* 147
rhizomorph 207
*Rhizopus* sp. 3, 178
  R. arrhizus 14, 180
  R. nigricans 121, 180
rice 12, 153, 155
root-infecting fungi 157–9, 161, 162
Rootsheild® 164
roquefortine 173
roridins 158
rosanes 94–6
rosein III 13
rosenonolactone 16, 22, 95–6
rosololactone 95
Roux bottle 207
rubrofusarin 130, *131*

rubroglaucin 129
rubrotoxins 125, *126*
*Russula* sp. 90
rust fungi 207

*Saccharomyces cereviseae* 8, 105
*Saccharopolyspora erythracea* 17
*Saccharopolyspora erythraea* 184
St Antony's Fire 165, 166
Sandimmum® 46
saprophyte 2, 207
sativene 85
*Schizophyllum commune* 26
sclererythin 7
*Scleroderma citrinum* 134
sclerotium 207
*Scolytus multistriatus* 159
*Scopulariopsis brevicaulis* 9, 27
secalonic acid 166
seiridin 163
*Seiridium* sp. 163
septate hyphae 207
*Serpula lacrymans* 71, 133, 142
sesquiterpenoids 73, **76–96**
  botryanes 81–4
  culmorin and helminthosporin 84–5
  cyclonerodiol 77
  helicobasidin 78
  metabolites of Basidiomycetes 85–93
  PR-Toxin 81
  trichothecenes 78–81
sesterterpenoids 73, **104–5**
*Setosphaeria khartoumensis* 126
Shikimate pathway 31
  pigments from 132–5
shikimic acid 31, 132
sicannin 116, *117*
silvathione 42
silver-leaf disease 162
simstatin 67
smut fungi 207
solanopyrones 151–2
solvent extraction 22
sorokinianin 156
spectroscopy and structure elucidation
  13

*Sphaceloma manihoticola* 97
spinulosin 52
spores 3, 148, 207
sporidesmins 42
squalene 105, *106*, *107*
squalestatins 126
*Stachybotrys* sp. 80
  *S. atra* 167
  *S. chartarum* 167
stachybotrytoxicosis 167
staggers, paspalum 166
*Staphylococcus aureus* 6, 10, 11
statins 14, **66–7**
*Stemphyllium radicinum* 16
*Stereum hirsutum* 91
*Stereum purpureum* 89
steroids
  fungal *see* triterpenoids
  microbiological hydroxylation 180–1
sterpuric acid 89, *90*
stictic acid 140, *141*
stink horn 142, 145
stipe 207
stipitatic acid 56
stipitatonic acid 56
*Streptomyces* sp. 207
  *S. cellulosum* 184
  *S. clavuligenus* 11, 35
*Streptomycetes* sp. 11, 184
streptomycin 11
strobilurins 14, 137, *138*
*Strobilurus tenacellus* 14
structure
  of fungi 2–3
  metabolite elucidation 13
Sudden Oak Death 162
*Suillus grevillei* 132
sulfate ions 26–7
sulfur, odours containing 144–6
sulfur tuft 115, 137
symbiosis 1, 2, 207

T-2 toxin 158, 168
taxonomy 4–5
tenuazonic acid 153
terpenoids 22, **73–119**

biosynthesis 73–6
biotransformation studies 184–7
defined and described 73, 207
diterpenoids 93–103
meroterpenoids 57, 73, 116–19
monoterpenoids 76
pigments 138–40
sesquiterpenoids 76–93
sesterterpenoids 104–5
triterpenoids and steroids 105–16
terphenyl pigments 132–3
terrein 60–1
tetrachloro-1,4-dimethoxybenzene 143
tetracyclines 11, 36
tetraketides 51–8
  gladiolic acid and derivatives 55–6
  6-methylsalicylic acid 51–2
  mycophenolic acid 57–8
  patulin and penicillic acid 52–5
  tropolones 56
2,4,5,7-tetrathiooctane 144
tetronic acids 122–3
thallus 3, 208
thelephoric acid 9, 133
thiamulin® 14
*Tinea pedis* 20
toadstool 3
tobacco anthracnose 155
*Tolypocladium inflatum* 46, 184
toxin *see* mycotoxin; phytotoxin
trace metal solution *19*, 20
tree diseases 159–63
tremulenolide A 91
triacetic acid lactone 50
*Tribolum confusum* 91
2,4,6-trichloroanisole 27
2,4,6-trichlorophenol 27
*Trichoderma* sp. 6, 80, 146
  anti-fungal activity 163–4
  *T. harzianum* 45, 146, 164
  *T. viride* 40, 41, 45, 111
*Tricholoma aurantium* 134
*Tricholoma flavovirens* 132
tricholomic acid 174, *175*
*Trichophyton mentagrophytes* 165
*Trichophyton rubrum* 165

trichorzianines 164
trichothecenes 13, 15, **78–81**, 158
  as mycotoxins 166–8
trichothecin 16, 22, *25*, 95, 167, *168*
*Trichothecium roseum* 24, 78, 94, 96, 167
tricyclazole 5
triketides 50–1
trimethylarsine 27
trisporic acids 139
triterpenoids and steroids 5, 73, **105–16**
  Basidiomycetes triterpenoids 113–16
  ergosterol 106–7
  fusidane steroidal antibiotics 107–11
  viridin and wortmannin 111–13
trochothecin 158
tropolones, tetraketide 56
tropophase 23, *24*, 208
truffles 145
tryptophan derivatives **42–4**
*Tuber magnatum* 145
Turkey X disease 169
tyrosine 5, *6*

*Usnea* sp. 141
usnic acid 141

vacuole 208
velleral 90
velutinal 90
verrucarins 158
*Verticillium fungicola* 143

*Verticillium hemipterigenum* 184
Victoria blight 156
victorin 45
victoxinine 156
virescenosides 94
viridicatin 42, *43*
viridicatol 42, *43*
viridin 111–13
virone 113
*Vitis vinifera* 150
volva 208
vomitoxin 167, 168
vulpinic acid 141, *142*

wild fungi 28, 208
wortmannin 111–13
wyerone 150, 151

xanthones 130–1, 141
*Xanthora* sp. 141
xenobiotic transformations 177–83
xenovulene 86, *87*
xerocomorubin 133, *134*
*Xerocomus badius* 28, 134
xylindein 131

yeasts 1, 2, 3, 5

zaragozic acids 126
zearalenone 169
zearanol 169
*Zygosporim mansonii* 68